Joe Martin in this series of graduation [...] *us to reflect deeply on the place of science a* [...] *our role as effective citizens of our various communities. Based on a long successful academic medical career, he offers helpful advice on the personal attributes for effective leadership. The reader will find the lessons learned from reading this book are insightful, profound and invaluable.*

—Frederick H Lovejoy Jr., M.D.,
William Berenberg Distinguished Professor of Pediatrics, Harvard Medical School

I know of no one better poised than Joe Martin to examine the interface between the certainties of fundamentalist religious thought and skepticism of the scientific method. Raised as a Mennonite, trained as a neuroscientist, challenged from all sides as a medical leader - including ten years as Dean of Harvard Medical School - he has a perspective that is truly unique. I have been deeply privileged to dialogue with Joe for many years - and now you can do so in the reading of this marvelous book.

—Dr. Tim Johnson,
Retired Medical Editor, ABC News and author of Finding God in the Questions

Few deans or university chancellors can informatively discuss the relevance of religious belief, moral duty, and responsible personal conduct as comfortably as they can medical practice or scientific research. Joseph Martin has done so. Drawing on his deep and unusually broad experience that span his youth as a Mennonite farmer in Alberta, to the upper-reaches of academic medicine, Reflections on Science, Religion, and Society, is both a guide and a provocative challenge to those who seek answers to current quandaries. His perspectives, spanning five decades, are crisp, specific, and practical.

—Hamilton Moses, III,
Professor of Neurology, Johns Hopkins School of Medicine

Like many of us, Joe Martin has struggled with the conflict between faith and science; in eloquent, literary prose, he welcomes his readers into his musings over these two contrasting and complementary dimensions of his life and career. He reminds us how remarkable it is that chemical reactions in the brain are the basis for the complexity and majesty of human intelligence—reflected in literature, love, altruism, benevolence, and the quest for knowledge, purpose, and meaning.

—Jules L. Dienstag, M.D.,
Carl W. Walter Professor of Medicine, Harvard Medical School

In this highly personal collection of commencement addresses and other commentaries, an eminent neuro-scientist/medical educator offers insights regarding the tension between facts and faith, with wise counsel for ways to address many of the significant sociopolitical issues of the day.

—Joseph Longacher, M.D.,
Former Clinical Professor of Medicine, Virginia Commonwealth University School of Medicine and Past President, Mennonite Medical Association

This collection of lectures, which span nearly four decades, represent the cumulated wisdom of one whose career embraced clinical medicine, front-line biomedical research, and senior administrative positions at two of the world's leading centres of learning – Harvard University and the University of California, San Francisco. Dr. Joseph B. Martin is eminently qualified to address the range of themes presented in these lectures as well as his present-day reflections on their relevance. Trained as both physician and scientist, his personal ethics, sense of integrity and caring concern for others were shaped by a nurturing family and rural community in Western Canada and by the tenets of his Mennonite faith. Dr. Martin's reflections will serve as guides for those who struggle to reconcile belief and science.

—Dr. Garth M. Bray,
Emeritus Professor, Neurology and Neurosurgery, Faculty of Medicine, McGill University

Reflections
on
Science,
Religion
and
Society

A MEDICAL PERSPECTIVE

JOSEPH B. MARTIN

◆ FriesenPress

Suite 300 - 990 Fort St
Victoria, BC, V8V 3K2
Canada

www.friesenpress.com

ISBN
978-1-5255-0489-1 (Hardcover)
978-1-5255-0490-7 (Paperback)
978-1-5255-0491-4 (eBook)

1. SCIENCE, PHILOSOPHY & SOCIAL ASPECTS

Distributed to the trade by The Ingram Book Company

Table of Contents

Book II. Medicine as a Profession

Book III. Varia

Further Reading

Acknowledgements

Two thousand and seventeen is the one-hundredth anniversary of two special parts of my history. In 1917, the Duchess Mennonite Church was established in a remote district of the Canadian prairies, founded by pioneer farmers who migrated from Pennsylvania toward the end of the First World War. In the same year, a group of Mennonite educational pioneers established Eastern Mennonite College, in Harrisonburg, Virginia, where my wife Rachel and I met in 1958.

Introduction

My father died in 2005 at the age of ninety-one. He was unable to accept the fundamental principles underlying Darwin's theory of evolution, specifically that over time, genetic changes could alter the form and function of an individual species. He would declare, "Show me when a monkey changed into a man!"

"But Dad," I would say, "How is it that the insecticides used in farming become ineffective with long-term use? That's an evolutionary change. And how come bacterial infections successfully treated by penicillin in the past no longer respond? That's the evolutionary development of resistance." Dad was a wise, widely read, intelligent, and curious person. His concept of the divine plan—the scriptural basis of the origin of our species—was rigidly interpreted. Sacred scripture overrode scientific insights.

He was like many of the church-going, deeply religious folks with whom I grew up. They accepted the literal interpretation of the King James Version of the Bible, and there was little room for intrusion of scientific discoveries. This presented a challenge for me as I embarked upon an appreciation of the methods and aims of scientific inquiry, where knowledge had a neutral valence. Medical practice, as I learned, is based on scientific credibility.

My mother, who died in 2014 at age ninety-nine, was a deeply spiritual person. Throughout her life, despite full support of my educational ambitions, she feared that I would lose my spiritual anchor.

As time passed, I found that the evolution in my perspectives of science and discovery were informed, and constrained to a degree, by my origins, so that the latter framed and influenced my perceptions.

What does faith offer in the world of scientific exploration?

How does scientific discovery influence what we believe?

How does what we believe affect what we do?

Over time, I learned that many of the scholars, scientists, and educators that I encountered and respected were of an atheistic (or disinterested agnostic) bent. I also grew to appreciate that although scientific exploration provides extraordinary information about the *how* of our existence, it offers little about the *why*.

As Ralph Waldo Emerson so eloquently said, "Life is a journey, not a destination." The speeches in this monograph reflect some of my thoughts, my mindset, and my musings about science, medicine, and religion during my journey. I have presented my story and perspective to various audiences over a thirty-five-year period, and reproducing them here causes some inevitable repetitiveness and redundancy, which I hope readers will forgive.

Prologue

It is more blessed to ask forgiveness than permission.
—Jesuit credo guiding saints and sinners
trying to find the right way to live

The world is full of places; why is it that I am here?
—Wendell Berry, American poet

In my memoir, *Alfalfa to Ivy*, I trace my journey from farm boy to a career in medical research and academia, with reflections on the influence of the conservative religious community into which I was born. In this series of speeches, I share some of my experiences with, and thoughts about, reconciling such radical notions as evolution with the tenets of my Mennonite upbringing.

Beyond the issue of career diversion—from my early aspiration to be a medical missionary to the appointment of major administrative responsibility at Harvard Medical School—I have maneuvered through the intellectual challenges created when my background and emerging appreciation of observational science have been in conflict. This includes a universe that is fourteen billion years old and the scientific underpinnings of genetics and neuroscience.

These issues dissect the human experience with reconsideration of nature versus nurture, free will, biomedical ethics, the meaning of empathy and good deeds, and, ultimately, the meaning of good and evil. They can lead us to ponder deep questions: Who are we, why are we here, to whom should we listen, and what brings satisfaction and happiness to our lives? These concerns have provided me with the fabric for commentaries and essays delivered in various settings, frequently commencement addresses at Mennonite colleges and seminaries, as well as graduation exercises at prominent medical schools.

In the first half of the book, I reflect on the interface of science and religion and the power of action in a world deeply in need of inspiration and focus. In the second half, I reminisce about various aspects of the medical profession. I remain passionately devoted to the notion that medicine is the greatest of all professions, but it is beleaguered by the forces of change imposed by modern complexities of organization, technical advances, and cost, each of which is having profound impacts on the delivery of equitable healthcare. Moreover, there are challenges inherent in the genomic revolution, which seems capable of evoking a new meaning for the "evolution" of our species.

In the final chapters, I discuss issues in the consideration of affirmative action in academic settings and some of the joys and challenges of academic leadership.

Book I. Belief, Science, and Ethics

CHAPTER 1

Commencement Address, Eastern Mennonite College

May 24, 1981

In 1958, I took a break from medical school at the University of Alberta and spent a year studying at Eastern Mennonite College (now University) in Harrisonburg, Virginia. It was a year that changed my life. Shortly after arriving at EMC, I met Rachel Wenger, who would become my wife two years later. We have now been happily married for nearly sixty years. It was the most important decision I have ever made.

When I gave this speech in the spring of 1981, I was the Bullard Professor of Neurology at Harvard Medical School and the chief of the neurology service at Massachusetts General Hospital. Using recombinant DNA was quite new in medical laboratories doing genetic research, but it opened up possibilities that raised a number of ethical issues, which I touch on in my speech.

Energy was a major topic in the news at the time. Issues of concern included the Three Mile Island nuclear accident in Pennsylvania, the worst in U.S. history, and the actions being taken by the Organization of the Petroleum Exporting Countries (OPEC), both of which I mention. Decisions taken by OPEC significantly contributed to the high gas prices and fuel shortages Americans experienced following the 1979 oil crisis, precipitated by the Iranian Revolution that year.

While I had spoken at Eastern Mennonite University before, I had never given a commencement speech. I was conscious that I was addressing graduates who had idealistic visions of the role they would have in the years ahead.

The Burden of Integrity

President Detweiler, Class President Samuel Augsburger, class of '81, alumni, parents, friends, and guests:

I must confess that I was pleased when Dean Keim called a few months ago to invite me to address the graduating class of 1981. The personal satisfaction was, however, short-lived. Several Mennonite friends and relatives, including my in-laws, were heard to comment, "But I thought only a preacher does that." Despite their misgivings, I'm delighted and honored to be here. It's really great to be back.

I assume that graduating classes at EMC are unlike those at Harvard. Former Harvard President Nathan Pusey was asked why the school is such a fount of knowledge. "Why that's simple. Each year, the freshmen bring in so much and the seniors take out so little."

I remember clearly the mixed emotions that flooded my consciousness as I sat in one of these chairs twenty-two years ago—as I pondered the future through perceptions and memories of the past—eager to engage in, yet uncertain about, the nature of the mission ahead. There are 268 of you in the graduating class today. I can guess that your backgrounds are more homogenous than your futures will ever be. You will go from here to a variety of occupations, positions, and endeavors.

Where will you be in five years, or ten? It is likely that many of you will find yourselves in some form of work concerned with the needs of others—in teaching, nursing, counselling, or social work. Some of you will embark upon graduate education in the arts or in science, or enter theology, medicine, or law. Others will pursue opportunities in business or farming, joining, in some cases, well-established family enterprises. And some will venture into areas of technology, such as engineering or computer programming.

The challenge ahead

Whatever you choose to do, I challenge you today not to underestimate or to forget a particular burden that has been placed upon you. It was placed there by the circumstances of your birth and your upbringing, and by your education within these halls at Eastern Mennonite College.

I call this the burden of integrity. Integrity has several meanings. In its simplest form, it entails honesty, trustworthiness, and incorruptibility. In its deeper sense, it means wholeness, completeness, and unity.

We were taught as children the full significance of its meaning. It was instilled into us in Sunday school. You have witnessed its definition, both explicitly and quietly, in your experiences here at EMC.

Integrity gives a particular depth to the meaning of equality, to the meaning of caring, and of sharing, a meaning that remains too unique in our world. Even though you may try to escape it or to dilute it, I can assure you that integrity has made its imprint and that you can never fully be content with your life unless you remain true to its burden.

Even though you may try to escape it or to dilute it, I can assure you that integrity has made its imprint and that you can never fully be content with your life unless you remain true to its burden.

Aerial view looking west of EMC 1952

Inquisitive intelligence

A college education achieved within the context of a Christian community brings together two great human ventures. The first views the world with a focus on inquisitive intelligence. It values accurate, testable knowledge. It experiences the sheer joy of knowing, of understanding the world, and of making discoveries. It asks questions and seeks answers. It enjoys the power that comes from an ability to predict and to control. This is the venture of science. It is also the venture of the social sciences and the humanities.

The second venture meets the world and the universe in wonder, trust, and commitment. It values the relationship of persons to each other and to their ultimate origins and destiny. It glories in the beauty of holiness and the responsibility of service. This is the venture of faith.

For those with an Anabaptist Mennonite heritage, the venture of faith is joined to a particular vision. I don't need to describe to you the particular genius of the Anabaptist insight, of the special meaning given to the equality of each individual, of the call to discipleship, or of our mission of peace. I can tell you that the world needs to hear about these things and that you should not hesitate to wear your chosen symbols openly and forthrightly. We are known because we care; we are recognized because we share; we are at times despised because we establish boundaries and limits.[1, 2]

We are known because we care; we are recognized because we share; we are at times despised because we establish boundaries and limits.

As one views the challenges of the eighties, it becomes evident that we will need to address certain ethical and moral

---

1   My Uncle, Sam Martin, was denied classification as a conscientious objector and served eighteen months in prison in Canada during WWII.

2   Influential in my upbringing was "Martyr's Mirror," written by Dutch writer Thieleman J. van Braght, a 17th century account of Anabaptist martyrdom.

problems that we have not faced before. There is concern at large about the capacity of our society to address these issues. It is also clear, in my opinion, that the moral majority will not provide a solution.

The Rockefeller Foundation, in 1978, formed a commission on the humanities to study the role and functions of this branch of human endeavor. I quote briefly from this report:

> *Through the humanities we reflect on the fundamental question: what does it mean to be human? The humanities offer clues but never a complete answer. They reveal how people have tried to make moral, spiritual, and intellectual sense of a world in which irrationality, despair, loneliness, and death are as conspicuous as birth, friendship, hope, and reason.*
>
> *By awakening a sense of what it might be like to be someone else or to live in another time or culture, they tell us about ourselves, stretch our imagination, and enrich our experience. They increase our distinctively human potential.*

The commission report goes on to say, "We are deeply concerned about serious social deficiencies of perception and morale. Our society has increasingly assumed the infallibility of specialists, the necessity of regulating human activity, and the virtues of human consumption. These attitudes limit our potential to grow individually and to decide together what is for the common good."

At the highest level, the humanities provide a foundation for exploring and gaining insight into the human condition. They do not, however, adequately address the issues of why we are here, where we are going, and in particular, say little about how we get there.

You view the human experience within a particular framework of divine indulgence and Christian commitment, where the meaning of humanity implies an individual worth that supersedes societal needs. Yet the individual is indelibly yoked to the problems of society and must contribute to the solutions. Our highest ideals are *humanity plus*.

Christianity meets technology

I want to emphasize a few issues that will require crucial decision-making within society at large, and within each of our own experiences in the decades ahead. As Christian citizens, we will need to address important issues that arise in a technological society. We must accept some responsibility to participate in the decisions that will be made, which will affect both us and our children. Your education, together with your integrity, provides you with special tools to approach the solution to these problems.

What are some of the issues that currently face us?

Over the past few decades, we have seen a distinct shift in the focus of our concerns. In the 1960s and '70s, the chief issues were social justice, equal rights for the races, and an obsession with the exploitation of under-developed countries by the military-industrial complex that this country has spawned. These are still important issues.

However, Three Mile Island, OPEC, ecological concerns, and technological advances now occupy much of our thought. In my view, it is the latter—the advances in science and technology—that threaten not only to change our ways of living, but our understanding of life itself. This will become a major issue in the last part of this century.

We have an enormous dependence upon technology in our society. We expect to have immediate access to the most recent information about the events that transpire in our world—the latest newspaper in the morning and the TV in the evening. The computer and its store of information have invaded our lives. You can't survive these days without a credit card, and you can't get one without someone, somewhere, preparing a dossier, which is then shared with others for the rest of your life.

The computer and its store of information have invaded our lives. You can't survive these days without a credit card, and you can't get one without someone, somewhere, preparing a dossier, which is then shared with others for the rest of your life.

I remember an incident a few years ago while trying to rent a car in Boston from Hertz without having made any prior arrangement. I was living in Montreal at the time and had no U.S.-based credit card. It took forty-five minutes for their computer to search each and all of the J. Martins whose credit was bad—some twenty-seven in total, as I remember, from all over the country—before I was cleared and given a car.

Within a few years, a majority of homes will have their own computers; many do now. Even the Sears catalog will become available this fall on video discs, which can be plugged into our television sets. Technology will, in the years ahead, reach close to the fundamental principles that govern life and thought. The human genetic material contained within our chromosomes will yield much of its secrets in the next two decades. It is estimated that the total information contained within human DNA would fill all of the pages in fifteen sets of the Encyclopedia Brittanica. To date, about 1% of human genes have been mapped.

I participated in a workshop at Woods Hole on Cape Cod two weeks ago in which a group of molecular biologists met with a group of brain scientists to discuss future research strategies. It was stated that by the year 2000, the whole of the human genetic material will be mapped and the information stored in computerized gene libraries. Almost certainly most of the human genetic hereditary disorders, of which there are currently known to be about three thousand, will have been placed into this map, and the ability to detect and determine the affected offspring charted for each. As one scientist said, "We may soon discover immortality, but it will take forever to prove it."

Genetic conundrum

Following closely upon this newer understanding will be techniques to make genetic corrections, glibly referred to as genetic engineering. It may become possible to correct enzyme defects that lead to mental retardation or to premature death from other causes. The cost will be large, and only a few may have the means to request and to get it.

Work in our department at Massachusetts General Hospital is focusing on one hereditary neurological disorder called Huntington's chorea, the disorder that claimed the life of Woody Guthrie, the folk singer. The disease is autosomal dominant, which means that each child of an affected parent has a 50% chance of developing it. What is particularly tragic about the disorder is the fact that the symptoms of uncontrolled movements, called chorea, and of mental decline or dementia, occur usually only after the age of thirty-five and into the mid-forties, by which time children have been born. There are no techniques presently available to diagnose this genetic disorder. The available techniques of molecular genetics, together with an analysis of the genes in such individuals, make it almost certain that a pre-symptomatic detection test will be developed within the next five years. Would you want to know that you were going to develop the symptoms of an untreatable disease? Would you have your children tested before they have children to break the chain? Would you think society would be justified in requiring everyone to have the test, just as we now require premarital blood tests?

The techniques used for this work are referred to as recombinant DNA. These methods provide the means to mix cellular genes from one species with those of another. In searching for the specific localization of a human gene, it is necessary to fuse human cells with mouse cells so that the genetic material coalesces. The cells can then be allowed to divide, with the genes from the mouse contained within the human cell as a permanent cellular marker. The biomedical ethical implications of experiments that alter the genetic structure of DNA readily become apparent. Such ethical issues, in my view, will outweigh concerns over the safety of these experiments.

The business world has already recognized the importance. Within the past two years, more than thirty companies have been formed to develop and market the products of molecular biology and molecular genetics.

In medicine, there are already several pertinent questions. Serious, even fatal, genetic defects can now be detected during pregnancy. Should a sick, maimed, or developmentally disabled individual be allowed to be born? Many say not.

Unpacking the brain

The other technological explosion that has emerged only in the past five to ten years is our understanding of the brain and its functions. The human brain contains more than ten billion nerve cells interconnected with each other. Chemical signals between these cells determine brain function. We have learned that the brain makes many chemicals that have been discovered and used for treatment for centuries. For example, recent discoveries have shown that the brain produces substances that act like morphine or heroin. They act normally to modify our perceptions of pain by regulating communication between nerve cells.

As doctors, we administer morphine to alleviate pain; it appears that the natural substances within our brain act in the same way as the materials found in plants. They also affect mood and other behaviors and, of course, carry great risk of inducing addiction. The point is that the brain uses such chemicals in a natural state. We know that it also contains local receptors that account for the anti-anxiety effects of valium and the tranquilizing effects of the drugs used in schizophrenia and in the manic-depressive [bipolar] diseases. It is the challenge of modern neuropsychiatry that mood, and perhaps of thought and personality, resides within these chemical pathways.

As doctors, we have reached an advanced stage of technology where patient treatment is sometimes determined by financial issues. Let me give one example. More than twenty thousand people are alive in the world today who, at any other time, would have been dead. They have lost their kidneys and are kept alive by the artificial kidney—a machine. In 1960, there were only a few such machines in the world, but there were thousands of potential patients. Who were to be the fortunate, chosen few to be permitted an extra bit of life, and who would be the unfortunate ones allowed to die? Committees were formed to make the agonizing judgments. Should a brilliant scientist without a family be spared, or a laborer breadwinner of a large family? Fortunately, the problem of selection no longer exists, but the cost of life may be as high as $20,000 per patient per year.

Kidney transplants can minimize that cost, but there are not enough organs to transplant. Most of them come from patients declared brain dead but whose kidneys still function. Decisions have to be made about stopping those vital functions that remain to permit removal of the kidneys for transplantation. Medicine has achieved a degree of technology and expense that threatens to limit the total resources available to our society. The Massachusetts General Hospital, long recognized as a leader in matters pertaining to health care and policy, determined recently that it was unwise to initiate a heart transplant program. Although the techniques are now relatively straightforward, the problems of case selection and the extent to which available resources would have to be reallocated, from many to a few, led to the decision not to initiate the costly program.

You know, health care has always been a problem. Plato had two criticisms of medicine. The first was that doctors treated slaves as carefully as they treated free men or philosophers, which did not accord with the rules of his authoritarian republic. The second was that doctors treated patients, including sick philosophers, as slaves!

We cannot neglect to mention a serious ecological issue facing us—our seemingly endless need for energy. Within this decade, it is likely that water will become *the* most important resource. I have often commented that the central issue most likely to lead to a major conflict between the USA and Canada will not be oil or gas or nuclear resources, but water.

These are some examples of issues, both current and anticipated. They are not necessarily problems that will be common to your experience. Nevertheless, whatever you do and wherever you go, you will feel the burden of addressing your integrity as it relates to the issues around you.

Hazards ahead

Let me turn finally to a brief consideration of some of the hazards that may lurk along the way for those who enter the world with a particularly heavy dose of integrity.

1. The first hazard, perhaps, is the tendency to arrogance, a flaw that can almost imperceptibly displace or misplace the trust that others would have in you. We view life from its highest ideal and claim to know the solution to many of its problems. However, we must confess that we do not know all nor have we as yet observed the many circumstances that render the view different to others. We must continually seek to listen and to become aware and knowledgeable about the views of others. It becomes important to test the reality and the workability of our hypotheses against the experiences that we gain from others. As Samuel Johnson said, "Integrity without knowledge is weak and useless, and knowledge without integrity is dangerous and dreadful." Be particularly wary of arrogance that is cloaked in humility.

Let us remember that we do not have an exclusive hold on the truth, and that we must learn from others. Your views will change and be molded by these vital experiences. It's a delicate balance to have an inner affirmation of that which provides meaning to life but to reserve judgment of others and to learn from them. Children help a lot when it comes to keeping us humble.

2. An equally serious hazard is the tendency to depression, discouragement, and despondency. Evil is all-pervasive in the world. It takes many forms. You may see serious failings in those you admire. The closer you get to people at the top of your particular ladder, the more evident their weaknesses become. It is easy to lose sight of one's goals because of the imperfections witnessed in others. Your attempts to address the wrongs can seem so trivial and unimportant. You are deluged with the noise and commotion of life, of assassinations, and individual crimes against personal integrity.

You may see serious failings in those you admire. The closer you get to people at the top of your particular ladder, the more evident their weaknesses become. It is easy to lose sight of one's goals because of the imperfections witnessed in others.

It can become a temptation to withdraw, to protect your own, and in that action to remove from the world the insight that you can provide.

The burden of integrity can become a burden of guilt; the burden of feeling driven to achieve something that actually makes a substantial contribution to the inequities of the world and a growing frustration about aspirations not achieved, of goals never realized. At this juncture, it is sometimes best to just live for a while and to take stock of the alternatives—maybe take a year off and go to Europe.

3. A third hazard follows closely. It is the remarkable tendency present in many of us to procrastinate, to wait for a better or ideal time to become involved. Jesus was thrown into the midst of the political and social events of his time and found himself on the side of the poor and against the formal organization of religion and of state. Each of you needs to develop your form of protest to the excesses of an insane world, to the military-industrial complex, and to a world that wastes energy and assumes little responsibility for the environment. Our society is now showing increasing tendencies to sequester wealth and to forget the poverty stricken. Big government is moving to lessen its burden of commitment to the poor and disadvantaged. That may be good, but it places responsibility upon you and me, and upon our private institutions and the church to pick up these responsibilities. None of us has clear answers to these problems, but they will not go away with procrastination.

4. A tendency to dissociate our faith and work. This may take several forms. On the one hand, it may lead to cultural isolation away from the realities of the world. Our ancestors sought this to escape unnecessary persecution; we must be certain that we do not seek it to avoid responsibility. On the other hand, there may be fear of participation with others who have similar but not identical opinions to our own. They may have similar general concerns but may have adopted different strategies.

There are those among you who will enter professions, disciplines, and occupations for which your background does not provide simple, reproducible, and unquestioned answers. These you will have to derive.

There is a need to recognize what I shall call the tension between our origins and our aspirations. I grew up in a Mennonite farming community in southern Alberta. I came, I think, from impeccable Anabaptist stock. My ancestors came to Pennsylvania from Germany in 1727. My aspirations have changed greatly over time. When I first aspired to become a doctor, it was to be a missionary, then a family practitioner, then a neurologist, and finally a university professor. None is better than the others, but the tasks are different. This change required a continual reassessment, reconsideration, and realignment of my origins and my aspirations. The move to major urban areas presents a concern for those of us who have taken such a course. How can we protect and transmit our heritage and faith to our children? How can they learn who they are? I believe strongly that one way is to encourage the kind of experience you have shared together at EMC. Perhaps in the language of the molecular biologist, we must learn to clone integrity.

Above all, let me encourage you to be yourselves. Don't try to copy others. There is no substitute for the ability to look another directly in the eye, both in times of harmony and in times of disagreement. It has been of particular interest to me to observe how insecure and dependent those we most admire are on reassurance from others. We all need to be liked. The best assurance of this is to demonstrate openness and directness.

5. Finally, there is the hazard of taking oneself too seriously. Let me encourage you to take with you into the venture of life a full dose of humor. The best protection against the follies of life is a robust sense of humor, richly laced with a realistic view of oneself. There is particular beauty in the ability to sense the ridiculous and to be able to laugh at oneself.

The best protection against the follies of life is a robust sense of humor, richly laced with a realistic view of oneself. There is particular beauty in the ability to sense the ridiculous and to be able to laugh at oneself.

I have found as I travel and lecture that I am frequently introduced with the note that I obtained my undergraduate degree from Eastern Mennonite College. This has given me a wonderful opportunity to tell my favorite ethnic joke. I know that some of you have heard it, but let me tell you anyway.

A Hindu, Jew, and Mennonite were traveling in the countryside of Pennsylvania when their car broke down. They approached the nearest farmhouse and asked to stay the night. The farmer was willing to oblige but stated that he had only two beds, so one of them would need to stay in the barn. The Hindu volunteered and went out for the night. A few minutes later there was a knock at the door, and the farmer, upon opening the door, discovered the Hindu standing there. "I am sorry," the Hindu said, "but I cannot sleep in the barn. There's a cow there." The Jew immediately volunteered to sleep in the barn and set off for the night. A few minutes later there was another knock at the door. Upon opening the door, the farmer found the Jew there. "I can't stay in the barn," said the Jew, "there's a pig there." The Mennonite then volunteered to sleep in the barn. A few minutes later there was a noise at the door. The farmer opened the door to discover the cow and the pig.

I call this the Mennonite antithesis of the ethnic joke.

Graduates, it's great to be alive and to be able to enter the arena at this point in history. You are blessed with a special set of talents and gifts. Go and use them.

Go forth in joy, in love, and in peace. And may God bless each and all of you.

Postscript

What a difference four decades makes! How inappropriate to give a thirty minute "sermon" in a commencement address. I have since learned that ten to twelve minutes is much preferred.

I predicted that most homes would have computers within a few years, but I could not have imagined the advent of smart devices, enabling people to carry computers with them virtually everywhere. In fact, the Internet was born in 1983. That was when a common protocol for interfacing with communications systems, first begun for military purposes through the actions of the Defense Advanced Research Projects Agency (DARPA) was made available to the public domain. I recall meeting some of the key individuals who participated in the planning of this initiative at the Institute of Medicine in the early nineties when I chaired a committee on mapping the human brain, where the potential for sharing large data sets was considered feasible through the burgeoning interconnections that we now call the Internet. At that time, the mapping effort was growing rapidly internationally.

When I reread this speech, one sentence stood out: "You can't survive these days without a credit card, and you can't get one without someone, somewhere, preparing a dossier, which is then shared with others the rest of your life." My concern about the collection of personal information was not unfounded, as evidenced by Edward Snowden's revelations concerning the level of data collection carried out by the National Security Agency.

I also referenced the ongoing work in my group related to Huntington's chorea, commenting, "There are no techniques presently available to diagnose the genetic disorder." Although it took longer than I predicted, this is no longer true. In 1983, my collaborators and I discovered that chromosome 4 houses the faulty gene. In 1986, we developed a genetic test to identify carriers of the gene. Some years later, in 1993, the gene itself was defined, leading to a DNA blood test that can reveal whether a person will be afflicted with this fatal disease. While this advance undoubtedly represents significant scientific progress, it also presents an emotional quandary for those genetically at risk who must decide whether to take the test. Prenatal testing can also be carried out, raising ethical issues around the termination of pregnancy.

Today we are realizing the completion of a dream finally realized in 2003, when the first complete human genome sequence was published in a few individuals at a cost of over a billion dollars. The effort to accomplish this extended over more than a decade. Now the price is approaching $1,000 for a complete personal genome sequence and can be accomplished in a few days. From this, we are witnessing an explosion of genes identified as relevant for understanding neurological and psychiatric disease. Sadly, no definitive, curative, or progression-delaying therapies for neurodegenerative diseases like Alzheimer's and Huntington's has resulted to date.

As we turn attention to the business of living, how, in practical terms, do we go about our daily lives? This is addressed in the next chapter.

CHAPTER 2

Commencement Address, Anabaptist Mennonite Biblical Seminary

May 19, 2006

Located in Elkhart, Indiana, the Anabaptist Mennonite Biblical Seminary (AMBS) was known as the Associated Mennonite Biblical Seminary until 2012. Students graduate with a master of divinity or master of arts degree, as well as certificates in various areas of study.

I gave this speech in the spring of 2006. I was in the ninth year of my ten-year tenure as dean of Harvard Medical School. It was my first visit to the seminary. While dean, I often had the privilege of being on a list of potential commencement speakers. But as dean of Harvard Medical School *and* a Mennonite, I received some invitations that would not likely be extended to others in my position. Such was the case with my invitation from AMBS. This was the first and only time that I spoke there. It was a special honor for me to address the graduates, faculty, and parents.

I mention two individuals whose work may be familiar to some, but not all, readers. Developmental psychologist and longtime Harvard Graduate School of Education Professor Howard Gardner is probably best known for his theory of multiple intelligences, introduced in the 1980s. Stated simply, it holds that people learn differently and should not be restricted to one way of learning. Another psychologist, Daniel Goleman, is a journalist and author of the 1995 book *Emotional Intelligence*. It details his belief that emotional intelligence, being aware of how our own emotions and those of others impact behavior, is as important to success in life as one's IQ.

Ethical Decisions that Transform Our Professional Lives: Leading by Listening

Honored graduates, distinguished faculty, spouses, parents, and friends:

I am so pleased to have been given the opportunity to attend your graduation ceremonies at AMBS.

Congratulations to each of you for earning the honor that comes to you today.

As I contemplated comments to make on an occasion like this, I found myself reflecting on the great biblical texts that have had an impact on my life; among those coming immediately to mind was the text from Psalms 8:4 in the King James Version.

"What is man, that thou art mindful of him? And the son of man, that thou visitest him?"

I like the simplified description of this passage in *The Children's Living Bible* from 1972[3]: "When I look up into the night skies and see the work of your fingers—the moon and stars you have made—I cannot understand how you can bother with mere puny man, to pay attention to him! And yet you have made him only a little lower than the angels, and placed a crown of glory and honor upon his head. You have put him in charge of everything you made; everything is put under his authority."

---

3   *The Children's Living Bible.* Published June 1972 by Tyndale House Publishers.

In the New International Version, it is rendered: *"What is mankind that you are mindful of him."*

With some poetic license and perhaps less than fully diligent exegesis, let me say how this passage impresses me. Two aspects compel discernment: First, "crowning his head with glory," and second, "mindful of him."

The two keys here for me are "head" and "mind," and I might add in a post-Renaissance, more scientific world—"brain" and "mind." I take this passage to mean that, as God's creation, we are crowned with intelligence, insight, curiosity, and a quest for meaning. We have been given a mind, presumably like God's, capable of contemplation and powerful enough to provide the tools for communication—the essence of humankind—through the ability to use language and symbols to converse with one another.

And so, I have entitled my talk today "Leading by Listening."

I can't imagine a more forceful charge to leadership than that. To learn by listening! I want to share with you some thoughts about the meaning of leadership from my own experience and suggest some ideas regarding their implications for the leadership roles that you have chosen.

Let me begin with a few autobiographical notes. I was born in Alberta, Canada. My father was a dairy farmer, and I was educated in a three-room school. In tenth grade, there were three grades in one room—eighth, ninth, and tenth—with one teacher, who was also the principal of the school.

I recall coming home from school with a report card and 95% on a test. My mother's response was, "What did you get wrong?"

Today is my mother's ninetieth birthday, and if I weren't here, I might be celebrating it with her in Alberta. She remains healthy and a continuing source of great wisdom in my life.

In grade eleven, I was bussed ten miles to a consolidated high school. There were twenty-one students in my graduation class, and I was very pleased to be the valedictorian!

Dreams and aspirations

My passion to be a doctor goes as far back as I can recall. I have a mental image of walking across a field when I was four or five and wanting to go to India or Africa to experience the kind of adventure that the missionary doctors told us about.

My family was part of the conservative Old Mennonite Church, whose families had migrated to Western Canada after World War I as pioneer farmers. They came primarily from Pennsylvania. My father was born in Hagerstown, Maryland, and my mother in Altoona, Pennsylvania. Both my parents were children when they arrived in Canada, where they met, married, and had four children. I am the eldest.

My beginnings at the University of Alberta at age sixteen were academically traumatic, but my choice to live with a Mennonite family in Edmonton, Alberta, was critically important to my early accomplishments.

Before graduation, my academic mentors encouraged me to specialize, and I chose Case-Western Reserve University in Cleveland to begin neurological training. Before that, another seminal event took place.

A serendipitous detour

After completion of the first year in medical school, I chose, with the dean's permission, to take a leave of absence to attend Eastern Mennonite College, as it was then called. That experience proved critical to my spiritual growth. I focused entirely on seminary-type courses. I learned some New Testament Greek, took Mennonite history from Irvin Horst, ethics with Chester K. Lehman, and choral conducting with J. Mark Stauffer. I graduated with a technically interesting degree that took into consideration my previous science background. They called it a BSc in Bible, and that's actually how it reads on my diploma. I sang in the touring male chorus and, most important of all, I met Rachel Wenger, to whom I have now been married for forty-six years. We compared notes of our ancestors soon after we met and discovered that members of our families, the Martins and the Wengers, had arrived together in Philadelphia on the same boat, the good ship *Molly*, in 1727.

At the end of my year at EMC, I was offered a teaching job at the college in the area of my major, chemistry. My father was blunt when I discussed my career options. "Why don't you finish medical school first?" And so I did.

After our wedding, Rachel and I returned to Edmonton, where I continued in medical school while simultaneously serving as assistant pastor and choir director at Holyrood Mennonite Church. Rachel attended the University of Alberta's School of Education to qualify as a teacher in Alberta.

Throughout our time together, Rachel and I have moved ten times and lived in many different cities, including Cleveland, Rochester, Hartford, Montreal, San Francisco, and Boston. In each experience, we benefited from fellowship with the emerging urban Mennonite churches and appreciated the continuing focus we felt there on justice, peacemaking, and integrity. But what we valued most of all was community—a fellowship and sharing of common stories and histories.

My research interests have been driven by a great curiosity about the brain and mind, about the "crown of glory and honor" that has been placed upon our heads—or indeed within our heads. My scientific writings recently have focused primarily on the neurodegenerative disorders Huntington's disease, Alzheimer's disease, Lou Gehrig's disease, and Parkinson's disease.

Similar callings

Your training has been similar to mine, with more emphasis in your case on spiritual development that leads to healthy minds and bodies. We occupy similar extraordinary positions from which to take on the task of dealing with the hopes and joys, the aspirations, and the pain of those we learn to know.

We are the principal professions: physician in my case, and pastor or counselor in yours, where the patient or church member expects to discover trust, compassion, and assurance of confidentiality. In both of our situations, there are few questions that we cannot ask, few barriers to open communication, and few obstructions to trust.

Both of our professions open up remarkable opportunities. An article in *The New York Times* a month or so ago emphasized how important seminary training is for preparing individuals for a variety of leadership roles. I am told that about two-thirds of your class expects to enter the ministerial profession. Others will undoubtedly contribute in a variety of ways.

Let me consider with you some of the attributes of leadership that may make your efforts to communicate and to lead by listening more successful. It has been said that God has given us two ears and one mouth so that we can spend twice as much time listening as speaking.

Five quotients that multiply to success

I want to talk about the attributes of truly effective leadership, and I am going to embellish my description of them by calling them "quotients," or measurements of leadership potential. In our community of believers, these are gifts of the spirit—that third, inscrutable dimension that accompanies brain and mind, making us truly the creation of God. I want to discuss five quotients. As we visit each of these, imagine how important they are, and will be, in your lives.

Quotient 1 – Intelligent Quotient (IQ)

Efforts to define a singular "g" for intelligence fill the pages of psychology and neurobiological journals and books. Howard Gardner at Harvard has described what he calls multiple intelligences.

By way of practical definition, I take intelligence to encompass the ability to learn, to remember, to synthesize, to create, to analyze, to differentiate, to classify according to type and condition, to construct new paradigms, to problem solve.

IQ implies ability to innovate, to think outside the box, and to construct new and novel scenarios.

Obviously, successful leaders have a distinguishing level of intelligence, but that alone is not sufficient without other attributes I wish now to consider. Individual brilliance may result in earth-shaking concepts, discoveries, and Nobel prizes, but of leaders we expect more.

Quotient 2 – Emotional Quotient (EQ)

This is the ability to understand another's position, to put oneself in the place and context of the whole, to empathize, to understand the impact of group dynamics on the outcome of a situation, to be able to reflect on one's own reactions, to feel and share another's disappointment and pain, to commiserate, and to plan in the context of the effect of an action on others. Simplified, it is the ability to listen, and to discern beneath the surface what the other person is *really* saying.

Daniel Goleman defines the competencies of emotional intelligence as self-awareness, self-management, empathy, and relationship skills. EQ includes sufficient temerity and curiosity to want to understand another's perspective. It includes wanting to learn from another in order to put right one's own views and impressions. EQ also implies an appreciation of self-insight that allows for dissent and disagreement without being personally threatened.

EQ is learning to lead by listening.

Quotient 3 – Humor Quotient (HQ)

This characterizes the capacity to see the humor, folly, foible, and ridiculousness of a situation. It encompasses the ability to use self-deprecation to accomplish an end, to exude a sense of lightness of being and charisma, and of good cheer and hope. It is the ability to detoxify a situation, to know how to relax the tension with a comment, a story, or a well-told joke. It is the ability to bounce back after an untoward event.

Included in this measurement is the ability to view things while hoping for the best possible outcome—a glass half full, not half empty. It is to feel good about oneself and the role one plays. It balances good will and good cheer with an appropriate balance of anxiety to set things on course and to toe the line toward an end. Included in the context of HQ is the ability to end a day with a sense of a job well done and with an ability to sleep well, to feel rested, and to prepare to take on the challenges of the next day.

It includes the ability to view life as a great adventure.

Doris Kearns Goodwin links humor to psychiatric approaches: "Modern psychiatry regards humor as probably the most mature and healthy means of adapting to melancholy."[4] George Valiant, a well-known psychiatrist,

---

4   Goodwin, Doris K. *Team of Rivals: The Political Genius of Abraham Lincoln*. Published 2012 by Simon & Schuster, NY.

has said, "Humor, like hope, permits one to focus upon and to bear what is too terrible to be borne." Quoting another unnamed source, Valiant stated, "Humor can be marvelously therapeutic. It can deflate without destroying; it can instruct while it entertains; it saves us from our pretensions; and it provides an outlet for feeling that expressed another way would be corrosive."

Quotient 4 – Judgment Quotient (JQ)

This is the ability to sum up a set of circumstances and know when to act, to know when the vectors are aligned to take the next step toward the end game. Judgment requires the ability to know when enough information is in hand to make the decision. It's the 80/20 rule: with 80% of the information in hand, it's time to act without worrying unduly about the other 20%.

JQ also applies to the principles of fairness, sometimes leading to decisions that are the best for the most, characterized by equity and equality when possible. Good judgment arises from integrity and honesty.

Quotient 5 – Generosity Quotient (GQ)

In many ways, a singularity of leadership success is epitomized in the term "vicarious living." Simply put, it is the joy and satisfaction that accompanies watching the success of others. In an organizational setting, as in a congregation or an effective office practice, it implies freely giving credit where credit is due, recognizing that there is no end to what can be accomplished if one does not care who gets the credit.

Another aspect to GQ is the ability to forgive and forget. Holding a grudge is a powerful disincentive to forward progress. It is impossible to hold a position of leadership without being the recipient of bad news that may reflect on one's own performance or of perceptions held of you as the leader. The source of these derogatory comments may come from important individuals whose roles in subsequent actions are critical. An effort to understand the context of the criticism is important when the account is first presented. Harboring negative feelings that come about from an inability to accept the criticism for its potential value can lead to persistent counterproductive relationships in future dealings with the individual.

Leadership in action

I want to conclude with some examples of leading by listening.

I am reminded of the challenges of communication in a demographically diverse society. Not only must we *accept* cultural differences; we must learn to *understand* them. When physicians treat patients who come from a cultural background different from their own, it may be necessary to appreciate that some of these patients will not share even the most basic belief about Western medicine—that more often than not, it works. I recall a magnificent book by the journalist Anne Fadiman, who illustrates this point with heart-wrenching clarity in her book titled *The Spirit Catches You and*

You Fall Down.[5] It details a "cross-cultural clash" between a Hmong refugee family from Laos living in California and the Western doctors treating their severely epileptic baby daughter Lia. The doctors recommended a complicated regimen of medication; the family refused, and Lia suffered irreparable brain damage.

At first, you may suppose this to be a clear-cut case of neglect on the part of the parents. If they had done what the doctors had asked, Lia would be fine. But what if I told you that the Hmong people believe—as fervently as you or I do in the science of medicine—that Lia's epileptic seizures were caused by demons who "caught her and made her fall down" and the only cure was to offer animal sacrifices to them. Lia's parents slept by her side every night and cherished her even after she had been returned to them in a coma. Who is to blame for this tragedy?

The doctors? They admitted Lia seventeen times even though her parents could not pay for her care. From this tragedy, one comes to appreciate how important it is to understand a patient's fears and doubts, no matter how ill-founded or strange they may seem.

Good advice for the ages: Don't assume

I tell our medical students to be cautious in assuming too quickly an understanding of the relationship between people and their diseases. What we accept today as dogma may be

---

5   Fadiman, Anne. *A Spirit Catches You and You Fall Down: A Hmong Child, Her American Doctors, and the Collision of Two Cultures.* Published April 24, 2012 by Farrar, Straus and Giroux.

wrong tomorrow. As we enter into therapeutic relationships with our patients, let it be more as partners than as authoritarian instruments. Some patients require a priest rather than a prescription. I have been puzzled by how the great physicians that each of us knows get their reputations. There seems almost to be a gift of healing instilled into some in extraordinary measure. I have come to believe that it is through the establishment of a unique doctor-friend-patient relationship that most of the healing occurs, whether helped along by surgery, acupuncture, or regular doses of approved medicines.

So what does it mean to listen?

It is by listening that the Mennonite Central Committee works so effectively. They stay focused on responding to the needs of local communities. The MCC currently works in sixteen African countries, including expenditures of $5.5 million in the Sudan.

It is by listening that Ten Thousand Villages meets with artisans in developing countries to provide them with half of the value of their products until they are sold in stores around the country. We have two stores in the Boston area, both doing substantial business. As recently described in *The New York Times Magazine*, the one hundred stores serve as the relief and development arm of the Mennonite Church.

It is by listening that John Hostetter, who was professor of sociology at Pennsylvania State University, assisted Victor McKusick of Johns Hopkins University in identifying genetic disorders in the Amish. As described recently in a book by

Susan Lindee,[6] such guidance was "absolutely essential for McKusick's genetic surveys of the Amish. To identify hereditary patterns, McKusick had to enroll a social network among the Amish in order to extend his medical knowledge beyond genetics to the local cultures and communities where the various maladies occurred." Dozens of rare recessively inherited diseases have been characterized, and genetic studies are underway to define potential treatments. Clinics now operate in the Lancaster area to aid the Amish in dealing with these conditions.

It is by listening that Mennonite Economic Development Associates (MEDA) provides microeconomic support to emerging family enterprises throughout Latin America, Africa, and Asia. MEDA is led by Allan Sauder, a Canadian and graduate of the University of Western Ontario. Sauder has lived and worked in the world of opportunities created by MEDA since 1987. He has traveled to more than sixty countries and has lived and worked in Tanzania and Bangladesh. MEDA successfully directs resources, principally from private philanthropy, as well as from federal government grants in the USA through the United States Agency for International Development (USAID), and in Canada through the Canadian International Development Agency (CIDA).

Let me share with you a story by Gabriel Garcia Marquez— *The Story of a Shipwrecked Sailor.*[7] In this poignant tale, nine

---

6   Lindee, M. Susan. *Moments of Truth in Genetic Medicine*. Baltimore: Published August 25, 2005 by The Johns Hopkins University Press.

7   Garcia Marquez, Gabriel. *The Story of a Shipwrecked Sailor*. Published March 13, 1989, Vintage.

Colombian sailors in a destroyer encounter a storm while returning home. Eight of the men are washed overboard; one survives. This lone sailor drifts in a small, half-inflated raft without food or water. Finally, after ten days, he washes ashore. Later, half-conscious, still lying in the sand, he is approached by a man who asks what has happened.

The sailor relates, "When I heard him speak, I realized that more than thirst, more than hunger, more than despair, what tormented me most was the need to tell someone what had happened to me."

All of your contacts throughout your career will have a story, their story, and it will be inextricably linked to their healing process—from illness, from poverty, or from neglect. Your ability to communicate, speaking and listening to those with similar stories and those with dissimilar stories, will determine in large measure your gift for healing.

Your ability to communicate, speaking and listening to those with similar stories and those with dissimilar stories, will determine in large measure your gift for healing.

As Barbara Brown Taylor said, "Life is a process of meeting strangers, meeting those that I have yet to learn to love."

In conclusion, I ask you to stand and repeat with me the prayer of St. Francis of Assisi. On a particularly difficult day, my chief of staff asked me what kept me calm and deliberate during difficult times. I went to my desk and showed her this prayer. It has been a source of great solace to me and I hope it will be also for you. It is a kind of Hippocratic Oath for service to humankind.

Lord, make me an instrument of thy peace.
Where there is hatred, let me sow love;
Where there is injury, pardon;
Where there is doubt, faith;
Where there is despair, hope;
Where there is darkness, light;
Where there is sadness, joy.

O divine Master, grant that I may not so much seek
To be consoled as to console,
To be understood as to understand,
To be loved as to love;
For it is in giving that we receive;
It is in pardoning that we are pardoned;
It is in dying to self that we are born to eternal life.

—Saint Francis of Assisi

Postscript

Today, brain scientists still ponder the relationship of mind and brain, of intelligence and free will versus determination and evolutionary bias.

During the initiation of movements, the brain anticipates the course of action before it occurs. "Mirror neurons" in the brain fire when observing the actions of another in the same area as the neurons activated by the actual movement performed. Our brains contain minds of comparison that might help explain feelings for, or empathy toward, others.

A great mystery, perhaps never to be fully resolved, is the nature of consciousness. Everything we observe, witness, and feel is an expression of our brain activity, and we know about the great mysteries of the universe because our brains are capable of conjuring up and studying the great questions. Perhaps two things will never be within the grasp of our understanding: How the universe is constructed and how our brains render intelligence sufficient to imagine how the universe works.

Each new scientific discovery—Einsteinian physics, the structure and function of DNA, the genetic basis of inheritance, quantum mechanics, and the brain wiring underlying fear, anger, and happiness—offer challenges to "simple biblical faith." Issues like these are taken up in the next chapters.

CHAPTER 3

Eastern Mennonite University Commencement Address

May 2, 2010

I gave my first commencement address at what was then Eastern Mennonite College in 1981, having received a degree there in 1959. Over the years, I have returned for various events and was honored to be invited back as the commencement speaker in 2010. In this speech, I refer to both EMC (Eastern Mennonite College) and EMU (Eastern Mennonite University).

After stepping down as dean of Harvard Medical School (HMS) in 2007, I focused on research related to Alzheimer's disease and Parkinson's disease. This work was carried out by my colleagues at HMS, where I was, and still am, a faculty member in the Department of Neurobiology.

My views on religion, like many things, have changed over the years, in large part due to my training as a scientist, but my religious upbringing continues to inform my interactions with others. Over the years, I have found that the most appropriate response is sometimes from the Bible. I remember being put on the spot during one particular meeting about the contentious merger of two Harvard-affiliated hospitals. I was asked what I would do to smooth the waters. I began by saying that I would "Rejoice with those who rejoice, and weep with those who weep," a passage I've quoted many times.

I believe that science, technology, and religion can coexist, and I value the principles of justice, peacemaking, and integrity that are central to the Mennonite faith.

Keeping Faith Relevant

President Swartzendruber, members of the faculty, honored graduates, family, and friends. Good afternoon. Thank you for inviting me to speak to you today.

Congratulations to each of you on this memorable occasion. Today, Eastern Mennonite University gives you leave to take on the next phase of your life's work, and I hope that work rewards you for the efforts you have made here to accomplish what we celebrate today. Despite what you may think are the best laid plans for your future, most of it is actually TBD—to be determined. Remember TBD; it provides a good excuse when pressed for answers about what you intend to do.

First, a couple of confessions: I grew up in a rural community in Alberta, Canada. In 1958, after completion of my first year in medical school, I surprised my dean, my family, and myself when I asked permission to leave medical school to attend Eastern Mennonite College, as it was then called. I only attended EMC for one year, but it transformed my life in innumerable ways—spiritually, emotionally, and philosophically. I came with a deep interest in learning more about my church, my faith, and what to do with my life. During the year, I focused entirely on seminary-type courses, such as ethics and Bible studies. I also learned some New Testament Greek— enough to read the Gospel of John. I took Mennonite history, learning for the first time about the origins of Anabaptist belief. I took music classes, choral conducting, and sang in the men's touring chorus. All in all, it was an incredible experience indelibly imprinted in my memory.

EMC was good to me. They added up my premed credits, counted medical school courses, and after one year here, I graduated with a BSc in Bible. That's actually how it reads on my diploma—B.Sc. in Bible, whatever that means!

Most important of all, I met Rachel Ann Wenger, from Columbiana, Ohio. In June of this year, just six weeks from now, we will have been married fifty years. Among the things we later discovered was that our ancestors, the Martins and the Wengers, had arrived together in Philadelphia aboard the same ship, the *Molly*, in September 1727, settling later within twenty-five miles of each other in Lancaster County. EMC brought two nomads, one from Canada and one from Ohio, back together again.

At the end of my year at EMC, I was offered a teaching job here to assist with labs and lectures in the field of my major, chemistry. I nearly did so, considering going on to seminary rather than returning to medical school, but my father suggested that I finish medical school first. And so I did.

Just a few decades later

My, how things have changed at this institution over the past fifty years. A college became a university. Undergraduate scholarship expanded and new majors were offered. New graduate school opportunities were created. New recognition came in collegiate sports—how proud we are of the Runnin' Royals.

A footnote: My grandson Gareth, a high school student, attends the early college program at Guilford College in Greensboro, North Carolina. My family was amused at the great pacifist double feature between the two schools—one Quaker, one Mennonite—that unfolded in March. The Royals won the first contest to enter the Final 16 but lost in the second. And look at the women's softball team, which was seeded eighth going into their tournament. And then they won! Now they will go on to play in the national Division III tournament for the first time in our history.

Aerial view of Eastern Mennonite University 2017
(Courtesy of Eastern Mennonite University)

But back to how things have changed here at EMU. Musical instruments are now played in the chapel. Special international recognition has come for programs in peace and justice; I saw in the graduation brochure that there are twenty-nine graduates in the Conflict Transformation Program, coming here to EMU from eleven different countries. Dates off campus are permitted without chaperones. And many new buildings and facilities have come into being through the great generosity of the university's friends and supporters.

Fortunately, some things remain the same. We are still nestled here in the beautiful Shenandoah Valley, with a majestic view of Massanutten. We have strong administrative leadership, determined to make the future here even better than the past. Notably, we still don't sing the national anthem before our sports contests!

Keeping up with the i generation

How the *world* has changed in fifty years. Slide rules gave way to computers. Telephones are superseded by email, iPods, iPhones, and now iPads. The only way to reach my grandchildren is by texting; they don't answer their phones anymore. Recently, I joined Facebook—the only way to keep up with family gossip—to which my son Neil responded, "Well, there goes the neighborhood."

Instant messaging keeps us distracted much of the day. Try holding a meeting today with everyone's Blackberry turned on. During lectures I give at HMS, the students read their emails, surf the web, and check out Twitter and Facebook, though they say they still hear everything I say. Who's kidding whom? You Skype with your family and friends back home. General Motors is in intensive care; Chrysler is moribund, teetering on life support.

But in 2010, some things have not changed, it seems. We still live in a world torn by war and violence. The U.S. seems ready at an instant to engage the enemy. We are always at the forefront in war, and our defense industry remains the most vigorous enterprise of the country, with constant growth supporting huge amounts of our national economy. In the 1960s, it was Vietnam; now it is Iraq and Afghanistan. We still struggle to understand the meaning of life. Science and technology have made religion more vulnerable to misunderstanding and misperception. How has that come about?

DNA unraveled

The human genome has now been revealed. More than twenty people have had their whole genome decoded, many placing information on the Internet for anyone to see. Techniques will soon be available to do yours, or those of your children, for less than $10,000, allowing us to view the weak spots encoded in our DNA, to predict disease, and explain mental and physical handicaps, and quite fearfully, to select the children we want to be born.

Genes are already known that predispose one to heart disease, cancer, and Alzheimer's disease. Medical care is being transformed by personalized medicine that reveals drug sensitivities. Brain imaging allows us to see what part of the brain we use to speak, to see, to hear, to feel, and to remember.

I am reminded of Psalms 8: "What is man that thou art mindful of him and the son of man that thou visitest him? Thou hast made him a little lower than the angels."

Here we are in 2010, where fundamental questions about life and death, about good and evil, and about how we relate to our fellow human beings and to this earth, are addressed. Despite knowledge from brain science, astronomy, and physics, the universe and the world order remain a mystery.

Each of us has in our creation in the image of God the capacity to learn about the vast resources of knowledge, so that we can apply it to our personal philosophy of living and giving. But:

+ How things will change in your lifetime is TBD

+ How will you respond to them, keeping your faith relevant?

In these last few minutes, I wish to reflect on three things that I have found important to sustaining my faith in the midst of unpredictable events and life experiences, as I have gained new knowledge and new appreciation for the complexities of our world, both the one internal to me—my thoughts and feelings—and to the one outside. Each comes from an expectation of gaining meaning through friendship and through our relationships with others.

Point 1. I call this "For where two or three are gathered together, there am I in the midst of them."[8] This promise has in my life unfolded in several ways.

After our wedding fifty years ago, Rachel and I returned to Edmonton, Alberta, Canada, where I completed medical school while serving as assistant pastor and choir director at Holyrood Mennonite Church.

Throughout our years together, Rachel and I have moved ten times and lived in many different cities, including Edmonton, Cleveland, Rochester, Hartford, Montreal, San Francisco, and Boston. In each experience, we benefited from fellowship with others of like mind in the emerging urban Mennonite churches. We came to appreciate in these experiences a continuing focus on justice, peacemaking, and integrity. But the biggest benefit of all was what each community provided—a fellowship where sharing common stories and histories could be used to ponder and attempt to address the great issues and questions of our time as our evolving experience and knowledge brought them to us.

I know that many of you do not come from a traditional Mennonite background, and I am thrilled to see all of you here. What I am suggesting is that the changes you will experience in your lives will benefit from the community of believers that you choose to be with. These relationships allow honest questions to be asked, challenges to your own perceptions to be raised, and an opportunity for growth.

---

8   Matthew 18:20 (KJV)

What I am suggesting is that the changes you will experience in your lives will benefit from the community of believers that you choose to be with. These relationships allow honest questions to be asked, challenges to your own perceptions to be raised, and an opportunity for growth.

Point 2. The second point may surprise you. I know that you are aware of the continuing tension between science and religion. Some of you have read Richard Dawkins' *The God Delusion*.[9] Others of you have struggled with the Darwinian explanation of life, even appreciating that it extends to religion itself, religion as a probable benefit of the tenets of natural selection for traits beneficial to survival of a tribe.

These debates continue as scientists explore the nature of the universe and as possibilities come into view that were never imagined before, requiring new reflections.

EMU has a great record of teaching science, whether in premedical preparation for medical school or in the education of nurses for their careers.

Science and religion can never be integrated into a whole singular frame of understanding. Science deals with testable hypotheses and observational data, presenting it as accurately as possible so that others can test to confirm or dissent from each observation.

---

9   Dawkins, Richard. *The God Delusion*. Published January 16, 2008 by Mariner Books. Richard Dawkins, a scientist and atheist, argues that belief in God is not only wrong, it can be dangerous, leading to wars and bigotry.

Religion, on the other hand, speaks to the meaning of life. It is concerned with the great issues of morality, justice, and equality, which are critical matters for the survival of our species. Driven by our religious nature, we ask questions that often have no answer, imponderable matters that depend on faith. We attempt to address our curiosity about the reasons for our existence, but these are matters that cannot be tested in a test tube, in a laboratory, or through a field trial. They are ideas that benefit from our talking to one another, learning from one another, and studying what others have said. The topic will never cease to be important as long as we have brains to contemplate these grand issues.

But fundamentally, we should not be threatened by scientific discovery. It will change how you view life. Look how electrical engineering has transformed our communication capacity through cell phones, computers, and the Internet. Even with respect to these transformations, science is inscrutable for most of us. I have no idea how three billion people can all be using invisible energy, radio waves, and magnetic fields that are critical to everything we do—and all be doing it at the same time. Tell me how that is possible.

Few of us now believe that the world is flat or that the sun orbits around our planet. Many of us no longer accept that our world came into existence in seven days six thousand years ago. Science seeks answers to these issues. We learn from science. We are obligated to and cannot avoid using what comes from its discoveries. But science cannot create or shape our faith.

Scientific discoveries will continue to shake our traditional understanding, and let me repeat, we ought not be threatened. There are questions that we think about that science can never answer. But we can learn from each other, we scientists and we religious believers.

> Scientific discoveries will continue to shake our traditional understanding, and let me repeat, we ought not be threatened.

Point 3. Now to my third point. How are we to live in the midst of this world that is our home and our future? I know no better way than the advice given by the prophet Micah: "But what does the Lord require of thee, but to do justice, love mercy, and walk humbly with thy God?"[10]

The world today cries out for the kind of sense of justice and wisdom that comes from the experience you've had here at EMU. How is that being realized today within our own Mennonite communities? I am proud that our church has developed ways to deal with these problems.

Let me suggest some of the initiatives that I have watched and admired in efforts by well-meaning people to lead the way in making the world a better, more compassionate place.

I am so pleased to learn of the activities of the Mennonite Central Committee (MCC), as volunteers labor in over sixty countries around the world—nearly a quarter of which are in Africa. These efforts include refugee assistance programs in

---

10   Micah 6:8 (KJV)

Iraq, Syria, and Lebanon; efforts to deal with poverty in Africa and Asia; and commitments to help build the infrastructure—economic systems that make healthy living possible.

I am amazed at the efforts in microfinance undertaken by the Mennonite Economic Development Associates.

In North America, the Mennonite Disaster Service sends teams of carpenters and tradesmen to devastated areas of the United States and Canada to deal with the aftermath of tornadoes, floods, and hurricanes. These are able men and women with organizational leadership and construction skills who are capable of helping survivors of disasters like those following Hurricanes Katrina and Sandy.

Beyond these examples, the MCC is made of everyday Christians serving in their local communities and neighborhoods, tending to the poverty stricken, and working as physicians and nurses in under-served regions of the country. I was astonished after a recent visit to MCC warehouses in Akron, Ohio, to see Amish volunteers working with conservative and more modern Mennonites in the common goal of preparing and sending off relief packages to all corners of the world. Another MCC initiative gathers volunteers to travel in mobile conveyances across the plains of the U.S. and Canada to prepare canned goods containing meat and vegetables that can be distributed in food packages to those in need.

These are the ways in which Micah's admonition can be realized in our time: "What does the Lord require of you? To do justly, to love mercy, and to walk humbly with your God."

I believe that the education you have been given here prepares you for this walk, and I wish you every blessing as you proceed from these halls and this campus into the next phase of your lives. Remember that you heard it here today: Your future is TBD—to be determined.

I close with the admonition made famous by Garrison Keillor, "Be well; do good work; and—from your alumni association—keep in touch."

God bless you all, and thank you for the opportunity to be with you today.

Postscript

Six years after the violence in Syria started, families are still displaced from their homes and struggling to meet their basic needs for food, shelter, work, and education. The instability of the Levant threatens all-out war with Syria, Russia, and Iran (now to a degree with Turkey) facing off against the United States, Israel, and Saudi Arabia. Sunni against Shia is a powerful undercurrent throughout. Despite such growing instabilities hope remains that global humanitarian efforts, many led by concerned religious organizations, will spare some from the tragedies of war, poverty, hunger, and homelessness. But such an outcome is by no means certain.

In the next chapter, the questions are asked, what makes us who we are? What motivates us to act and become involved?

CHAPTER 4

On Being Authentic

Convocation, University of Calgary
May 10, 2012

It is a tremendous honor for me to join you in Calgary for the graduation ceremonies for the class of 2012. I grew up on a farm a hundred miles east of here, along the old gravel road known as the Trans-Canada Highway, but Calgary was my home city. It provided big-city shopping, a zoo, and the pathway to Banff and the Canadian Rockies.

It is hard to believe that it was exactly fifty years ago this month that I received my doctor of medicine degree from the University of Alberta—at the time, the only medical school in the province. Today I don't feel much different until I check up on the guy in the mirror each morning.

A volume worth keeping

I brought along a book from my library, which I saved as my wife Rachel and I enter the debridement period of our lives, downsizing our belongings in a soon-anticipated move from our house to a more manageable condominium.

It is an important book to me. It details a research conference I attended here in 1973. I knew Calgary had opened a new medical school. I knew the first dean, Bill Cochrane, and the one who followed, Lionel McLeod. He was my mentor in medical school and encouraged me to pursue a career in teaching and research.

But back to the book. The title is *Recent Studies of Hypothalamic Function*, edited by two outstanding University of Calgary faculty members, Karl Lederis and Keith Cooper, both internationally acclaimed European scientists who recognized the amazing potential here and joined the faculty. They were harbingers of the great strengths in research that would follow. I had just embarked on a research career at McGill University, and my investigations focused on the brain regulation of the pituitary gland, particularly with respect to growth hormone. The conference was an outstanding compilation of the best work in the field in what was emerging as the new discipline of neuroendocrinology. I was impressed and observed with some envy the subsequent diligent recruitment of faculty to the University of Calgary Medical School, later supported by Peter Lougheed's great idea, establishing the Alberta Heritage Foundation for Medical Research.[11]

Calgary also soon became known for its radical curriculum reform, based on a systems approach to body structure and function in health and disease, and the successful demonstration that outstanding doctors could be educated in three years, not four. I cannot emphasize how devoted the faculty

---

11   [Peter Lougheed served as Premier of the province of Alberta from 1971 to 1985.]

were to the education of students, an engrained habit that persists admirably to this day.

Today we have another group of graduates to welcome, not only in graduate studies, medicine, and law, but including for the first time, veterinary medicine.

This reminds me of a true story. About fifteen years ago, I arrived late to a meeting of neurologists in San Francisco, checked into the hotel, and next morning joined a group of doctors at breakfast. I overheard them discussing the program for the meeting: epilepsy, diabetes, and so on. When the subject of herniated lumbar disks in dachshunds came up, I asked for a program. I had the wrong hotel. I was at the American Veterinary Medical Association meeting! I've often reflected on the great similarities between people medicine and animal husbandry, so I feel comfortable talking to all of you today.

The genesis of "you"

My talk is entitled "On Being Authentic." As you ponder the meaning of today and its impact on the life you are about to enter, you may find yourself reflecting as I have been doing. Looking back, I wonder how elements of character and personality develop from the template we are born with. How does experience imprint later characteristics that come to define us? How do sensory perceptions early in life affect who we think we are? How do these play out with the genes we inherit from our parents and grandparents?

Nature versus nurture. Likely not either, but rather both! For me, three elements of my roots influenced my life.

The first was early awareness of my family origins. This arose from family stories of generational migrations and the movements of our family, who escaped persecution in Europe to reach William Penn's Pennsylvania. Then western migration brought my grandparents from Pennsylvania to the expansive plains of Alberta, Canada. My father was a great storyteller, and his recollections of the Great Depression and of his hunting, fishing, and trapping adventures solidified my attachment to the soil, the mountains, and the Dinosaur Valley of the badlands near my home.

The second formative element was the realization that the happenstance of one's placement on the planet is a matter of extraordinary luck. As Wendell Berry put it, "The world is full of places; why is it that I am here?" Perceptions from everyday experiences embed biases about the cultural, societal, religious, and economic realities of that placement.

The third element that shaped my perspective relates to my origins in Canada, where values and sociological underpinnings had a powerful effect on my views of a citizen's rights and expectations. This was particularly so as it related to health care and the responsibility of government and society to work to serve the common good, without bias toward the wealthy. Starting early on in my academic career, this "socialist" perspective—now entrenched in the United States as a fundamental evil—gave me a desire to give back to my native country.

The third element that shaped my perspective relates to my origins in Canada, where values and sociological underpinnings had a powerful effect on my views of a citizen's rights and expectations. This was particularly so as it related to healthcare and the responsibility of government and society to work to serve the common good, without bias toward the wealthy.

O, Canada

I am grateful for my Canadian medical education. Not inconsequential in this regard was the low tuition expense I incurred, giving me the freedom to undertake prolonged postgraduate training without accruing large indebtedness, such a common issue now for graduates in the U.S.

When I returned to Canada, to Montreal, in 1970, I witnessed the implementation of Medicare, with universal healthcare coverage, financed by joint agreements between the federal and provincial governments. In Quebec, the government insurance scheme worked brilliantly. There was no concern about whether a patient had insurance, and the billing procedure was so simple and straightforward that an office clerk could manage every detail. I have no doubt that these experiences frame my present bias toward a universal healthcare system in the U.S., based on a single-payer system. However, I recognize that this is at present a political impossibility. The current system in the U.S. is broken, not only because of the uninsured, but because healthcare costs of the traditional fee-for-service reimbursement mechanism are unsustainable, and primary care networks essential for proper delivery of health care and for prevention are simply not in place. I have a sense of outrage at the social injustices that derive from the big business model of insurance and the healthcare administration now extant in the U.S.

The current system in the U.S. is broken, not only because of the uninsured, but because healthcare costs of the traditional fee-for-service reimbursement mechanism are unsustainable, and primary care networks essential for proper delivery of health care and for prevention are simply not in place.

Today is your day. You have labored long and hard to achieve this accomplishment. Congratulations. The world is your oyster; go ahead and create a few pearls. The professions you are entering offer so many opportunities. I urge you to follow the imprints that make you who you are, leavening your adventures with a commitment to make the world of medicine and other aspects of health care a better place.

About forty years ago I began to keep a journal. Not a diary, but a thought piece. I kept the reflections, and they became the framework for my recently published book, *Alfalfa to Ivy*.

In its pages, I reflect on four decades of academic experience, time spent in various leadership positions at McGill University; the Massachusetts General Hospital; as dean and then chancellor at the University of California, San Francisco; and lastly, as dean of Harvard Medical School.

My academic experience has spanned two countries, both public and private institutions, and healthcare systems organized with disparate philosophical backdrops.

Each step of my life has been heralded by unexpected opportunities, where serendipity played a major role. Today, looking back, I feel enormous gratitude for these experiences, resulting in a compelling drive to reflect on lessons learned along the voyage.

Keep a journal. You may want some day to write your story.

Bon voyage to all of you, and a note from your alumni association—keep in touch.

CHAPTER 5

Response to the 100th Annual William Belden Noble Lecture

Memorial Church, Harvard University
October 6, 1998

Free and open to the public, the annual William Belden Noble Lectures take place in Harvard's Memorial Church. Established in memory of a divinity student who died while preparing for the ministry, each lecture relates to some aspect of Christian thought. In keeping with the established format, a response to the topic covered is delivered immediately following each lecture.

Armand Nicholi, who was an associate clinical professor of psychiatry at Harvard Medical School in 1998, based his Noble lecture on the seminar he taught on Sigmund Freud and C. S. Lewis at Harvard College for more than thirty-five years, and later at Harvard Medical School. He titled his lecture "Sex, Love, and Joy: Contrasting Perspectives." In it, he explained that despite Freud's atheism and materialistic world view and Lewis' spirituality and embrace of Christianity, both men shared similar ideas about sexuality and monogamy.

Nicholi explained that Freud equated happiness with pleasure, including sexual pleasure, while Lewis believed that sexuality is not forbidden or sinful, but observed that most people struggle to stay monogamous or to abstain from sex.

In addition to Dr. Nicholi, at the beginning of this speech, I acknowledge Professor Gomes and Dr. Johnson. Peter J. Gomes was Plummer Professor of Christian Morals at Harvard Divinity School and Pusey Minister at the Memorial Church until his death in 2011. Timothy Johnson, MD, is best known for his work as a medical journalist with ABC News. He is also an ordained minister. Dr. Johnson moderated the 1998 Noble lecture discussion and was chosen to present the Noble lecture in 2004.

I also refer to Congress deliberating over impeachment. As many readers will remember, in December 1998, the House of Representatives impeached President Bill Clinton on two charges, perjury for lying under oath to a federal grand jury and obstruction of justice. Both charges were related to the cover-up of his relationship with White House intern Monica Lewinsky.

Sigmund Freud and C.S. Lewis—Either or Both?

Professor Gomes, Dr. Johnson, my colleague, Dr. Armand Nicholi. Thank you for the kind introduction and the opportunity to participate in this the 100th anniversary of the Noble lectures.

I don't believe that I have ever spoken in such a grand place. Indeed, it has been some time since I have spoken in church at all. So it is with considerable trepidation that I will proceed. I have, since my teenage years, been an admirer of C.S. Lewis, whose books I enjoyed reading in university as I made the transition from a small farming community, and a member of a conservative Mennonite church, to the big world of pre-medical and medical education.

I must confess that I have not read Freud extensively or deeply, although a part of my training as a neurologist involved time studying psychiatry, which I came to enjoy immensely. I chose to become a neurologist and a neuroscientist because I was fascinated with the question, "How does the brain work?"

But you must be careful about listening to what a neurologist has to say. I'm reminded of the tale of two men in a balloon over Boston:

> *As can happen, it was fogged in.*
> *They drifted aimlessly, and when the fog lifted, they*
> *were lost.*
> *Seeing a man standing in a field, they shouted, "Where*
> *are we?"*
> *The response: "In a balloon."*

"Just our luck we should get a neurologist," one of the men said.

Said the other, "How do you know he's a neurologist?"

"Easy. His information is perfectly accurate but completely useless."

One of my early struggles was how to deal with evolutionary principles and doctrine, coming as I did from a bibliocentric view of the world, but I soon discovered the "facts" regarding evolution to be so compelling that it was only a matter of time before those ideas were incorporated into my own personal world view.

In my years as a physician, I have witnessed a great deal of pain in my patients and, thankfully, of happiness as well. I have been greatly impressed by the resilience of the human spirit. I teach medical students that the physician ought to acknowledge that healing is a complex event, assisted by the physician's scientific knowledge and bedside skills, but that healing is also propelled by factors as yet poorly understood: the will to live, the hope of cure, a disdain of disability, and remarkable fortitude against all odds.

As a student of and believer in the reductionist approach to the study of the nervous system, I also recognize that as a physician, I must acknowledge that patients who believe in science as the foundation for medicine also know that science has not and probably never will provide the answers to deep-seated human attributes. These include happiness, a sense of well-being, and joy—or their opposites, depression and hopelessness.

Sex, love, and happiness

In my comments, I want to address the question of pleasure and the nature of sex, love, and happiness from three perspectives.

First, what can be learned from the perspective of evolutionary biology? Second, how has neurology and neuroscience informed us? And finally, what is my understanding of the Christian perspective on this issue?

I will give you the bottom line, so to speak, at the beginning. The validity of each of these perspectives has been demonstrated, but they cannot at this time be conjoined into a common theme or motif. So much, we might say, for consilience. Interestingly, when I typed that word into my computer, it sent a signal that it was misspelled, which I take to mean that it was not in the computer's vocabulary. That might not please Professor E. O. Wilson.[12]

Evolutionary Biology

The premise of Darwinian evolution is centered on natural selection through survival of the fittest. The physical attributes that emerge during selection are contained within the genetic material that provides both the physical characteristics that work to the advantage to reproduce in like kind and in the behavioral attributes that are genetically determined.

---

12   Wilson, Edward O. *Consilience: The Unity of Knowledge.* Published March 30, 1999, Vintage. Edward O. Wilson, a Pulitzer Prize–winning author, a scientist and proponent of sociobiology, and longtime Harvard professor, uses the obscure term consilience to describe the synthesis of knowledge from different fields. Wilson discusses methods that have been used to unite the natural sciences and contends that it is time for the humanities and social sciences to join with the natural sciences.

By the dogma of evolution, the strongest survive to reproduce and the weaker die. In our recent trip to the Galapagos, we were greatly impressed by the tameness of the birds and animals there, separated as they were during their evolutionary development from the Ecuadorian mainland six hundred miles away. They are in fact not tamed, but rather are fearless, having in most cases never been exposed to predators. So their genetic makeup and the behavioral consequences of their genes exclude fear compared to how it has developed in species more closely associated with our human-populated world.

In the crudest terms, reproductive fitness, or fecundity, reigns supreme in a purely Darwinian world. The survival of the individual is manifest in the production of offspring who carry the parent's genes into the next generation. The more fit the offspring, the better.

In a Darwinian world, sex, love, and happiness come down to sex—with that Freud would agree. Monogamy is disadvantageous in most circumstances. Here, gender differences emerge; the male has a distinct advantage in being able to impregnate many females, while the female is focused on successfully bringing the offspring to birth and maturity so the cycle can repeat itself. Recent studies show that the evolutionary drive, so to speak, commonly results in unfaithfulness even among animal species that have been considered monogamous, with both males and females reaching out beyond their partners for clandestine mating.[13] One might conclude that the pressure of genetic mechanisms is to unfaithfulness, a concept that may be of relevance in our times as Congress deliberates over impeachment.

Recent studies show that the evolutionary drive, so to speak, commonly results in unfaithfulness even among animal species that have been considered monogamous, with both males and females reaching out beyond their partners for clandestine mating.

In summary, evolutionary principles suggest that survival does not lead to faithfulness but creates a powerful incentive to move away from it. In terms of our evolutionary biases, sexual drive leads us toward sex as a pleasurable event, and it tends to turn us toward promiscuity. Sex feels good, and the more often it is performed, the greater likelihood of sending one's gene pool off into the next generation.

---

13   McGraw, L.A. and Young, L.J. *The prairie vole: an emerging organism for understanding the social brain.* Trends Neuroscience 33:103, 2010. Behavioral studies in burrowing rodents called voles show a remarkable biochemical difference between prairie voles, which are monogamous and meadow voles, which are promiscuous. Scientific observations relate this difference to the quantity of brain peptide receptors for two hypothalamic hormones, oxytocin and vasopressin. Prairie voles are found to have more receptors for these in the 'emotional' parts of the brain, which likely contributes to emotional bonding and monogamy.

Neurology and Neuroscience

It is quite evident that the evolution of the brain has changed little the anatomical localization of pleasure. It is possible to place an electrical stimulator into a specific region of the brain, and when current is applied, the animal can learn how to self-stimulate. By pushing a button, current can be delivered into that region. The effect is so powerful that animals cease virtually every other activity, including eating, to carry self-stimulation to exhaustion.

Now this is a complex experiment, and it has been argued that the repetitive stimulations are actually to prevent a bad sensation from occurring after each stimulation. But whatever the mechanism, such effects are sharply localized to the dopaminergic pathway in the brain, where the neurotransmitter dopamine is carried from one region of the brain to another. This is the same system that is activated by addicting drugs such as morphine. From studies of brain function, we can localize with some considerable precision the sites that define the fundamental drives. The brain is organized in a hierarchical manner. Let's examine what this means.

At the most primitive level, such drives exist within the hypothalamus, a small, walnut-size part of our brain located immediately behind the eyes in the midline. It is here that the four basic drives reside. These are the four F's, as they are often called: feeding, fighting, fleeing, and—SEX.

These drives are controlled or regulated by higher brain regions of the basal ganglia, amygdala, and cerebral cortex. One can identify regions that determine pleasure here, for example, where the brain is activated during states of addiction in animals or in humans. Sophisticated analyses of these locations by PET scans and MRI localize these functions precisely. In a further extension of the brain's hierarchy, these regions are under the control of the cerebral cortex, the so-called new, or 'neo'cortex. This region of the brain provides the anatomical substrate to locate Freud's superego and gives C.S. Lewis a place for free will!

In a further extension of the brain's hierarchy, these regions are under the control of the cerebral cortex, the so-called new, or 'neo'cortex. This region of the brain provides the anatomical substrate to locate Freud's superego and gives C.S. Lewis a place for free will!

Another comment about genes and feeling good or bad. We now know with considerable certainty that traits that predispose to mania and to depression are carried in our genes. Such bipolar disorders are in some families inherited as autosomal dominant characteristics transmitted directly from one generation to the next. The genes for these traits have not yet been found, but I have no doubt that they will be, indicating the power of our makeup in controlling our psyche. In a crude sense, our genes determine our drive to procreate, and our brains are wired to see that we do.

The Christian perspective

How does all of this connect to a Christian perspective? How do I bring together my science and my religion? I would ask the question analogous to the tree falling in the forest: Does it make a sound if there is no one there to hear it? The analogy in the human situation might be: If I were alone, would I know that I exist? We enter the world in contact with our parents, and we emerge from infanthood a product of our genes and our experience. Who we become is forged out of this complexity. As Robert Coles has so eloquently described, our perception and concepts of God appear early and probably universally.[14]

The Genesis account of creation describes explicitly this relationship: a man, a woman, and God.

To contemplate the meaning of joy or happiness or sex in the absence of human relationships is empty, vulgar, and meaningless.

I have been informed in this connection by the writings of Gordon Kaufman, one of this generation's most influential theologians, now a professor emeritus in the Divinity School here at Harvard. In his book, *In Face of Mystery*[15], he attempts to bridge the chasm between our biohistorical origins and our place in the world.

He writes that the characterization 'biohistorical' holds together and sums up reasonably well what many today, at

---

14   Coles, Robert. *The Moral Life of Children.* Published February 4, 2000 by Atlantic Monthly Press.

15   Kaufman, Gordon D. *In Face of Mystery.* Published February, 1993 by Harvard University Press.

least in the West, understand humans to be. Central to this notion is the contention that "humanity cannot be understood simply in and of itself, but only in relation to its context, the whole complex of life, which has gradually evolved on earth …"

He goes on later in the book to say, "Faith is essentially an ongoing struggle with the sin and evil in ourselves and our world, as we give ourselves over as fully as we can to that trajectory … which beckons us toward a more humane society in a well-ordered world."

How are we then to live? How do the elements of joy, love, and sex manifest themselves in our living? As Christians, our sexual mores evolve from the Golden Rule, treating one another as we would like to be treated, with respect for the dignity of the lover, the partner, the friend. These are uniquely human attributes. They are embraced by and expressed through acts of faithfulness, monogamy, and sex limited to love and commitment.

As Christians, our sexual mores evolve from the Golden Rule, treating one another as we would like to be treated, with respect for the dignity of the lover, the partner, the friend. These are uniquely human attributes. They are embraced by and expressed through acts of faithfulness, monogamy, and sex limited to love and commitment.

How should we interpret the great change in C.S. Lewis? As Armand Nicholi has described Lewis to us, he became a different man, turning from an inner to an outer perspective, from self-centeredness to generosity, from lonesomeness and loathing to friendship and sharing. Freud appears to have taken a different route. He turned increasingly inward. He became embittered toward his friends and removed himself from their camaraderie.

I believe that this informs us about the importance of relationships, of community, and of the human spirit at its best. It is in the context of such community that the fruits of the spirit emerge—love, joy, peace, patience, kindness, goodness, faithfulness, gentleness, and self-control.[16]

The first of these is love—agape love as we have learned tonight—and it is the basis of the two great commandments: to love your God with all your heart, and your neighbor as yourself.

This is part of the kingdom of God on earth.

---

16   Galatians 5:22-23.

CHAPTER 6

Sixth Annual George B. Murray, MD, Limbic Lecture

Department of Psychiatry, Massachusetts General Hospital
March 10, 2016

A few years ago, my wife Rachel and I viewed Michelangelo's paintings in the Sistine Chapel. The *Creation of Adam*, completed in 1512 and displayed at the center vault of the chapel, invokes the image of God enshrouded in a purple cape, the outline tracing a front-to-back (sagittal) section of the human brain. The frontal lobe is on the left, the posterior lobe and cerebellum on the right. The folds in the cortex are represented by cherubs, and God is placed in the motor activation zone in the center of the brain. God reaches out, as does Adam, but they don't quite touch. A gap—a synapse, if you like—exists between the God figure and the mortal. I imagine the twenty-first century as the *Century of the Synapse*, providing a construct for examination of brain structure, function, and disorder.

I explored this concept when I was invited to deliver the 6th Annual George B. Murray, MD, Limbic Lecture to the Department of Psychiatry Grand Rounds at the Massachusetts General Hospital (MGH). Dr. Murray was a colleague and friend during my years at MGH, where he was one of the leading psychiatrists in the emerging field of consultation-liaison psychiatry, lending wisdom on management of patients admitted to the hospital when behavioral problems arose.

Courtesy of www.michelangelo-gallery.org

My objective was to describe areas of biomedical inquiry ripe for the integration of neurology and psychiatry. Distinctions between neurology (where brain abnormalities are apparent on the whole brain or through microscopic examination) and psychiatry (historically based on a theory of disorders of the mind) are being conjoined by functional imaging, genetics, and growing recognition of overlap in symptomatology of prior diagnostic entities. This presentation provided an opportunity to deal with perceptions of the distinction, if any, between "brain" and "mind."

Brain and Mind: Visible and Indivisible

Dualism in philosophy, science, and religion

The concept of dualism as applied to the body and the soul, and to the "brain" and the "mind," dates back millennia, made famous by the writing of French philosopher René Descartes, who wrote passionately about their separability—*I think, therefore I am.*

Such dualism also occurs in pursuit of knowledge and understanding of the world we live in. It can be manifest in differing, often divisive, concepts in scientific, medical, and religious fields of inquiry. Today, I reflect on three examples of such experimental endeavors.

First, we will examine differences in concepts of brain structure and function as explored over a hundred years ago through the work of two European scientists, Camillo Golgi and Santiago Ramón y Cajal, contemporaries whose work spanned the late nineteenth and early twentieth centuries. The second focus will be on methods of study of brain function in health and disease, as perceived by Sigmund Freud and Emil Kraepelin, at the beginning of the twentieth century. As we shall see, these two examples remain relevant to our consideration today.

I will conclude with the most difficult perspective: How do we approach differences in perceptions of consciousness, belief, and the ultimate question, *Why are we here?* Is it an accident of nature, or the result of an intelligent design pointing to a creator and sustainer, God?

One method, two conclusions

In 1885, Golgi stumbled across a remarkable finding. When sections of brain were exposed to silver nitrate solution to yield a silver citrate stain, a small percentage of nerve cells stood out magically with full details of their cell bodies and extending processes of dendrites and axons. Golgi developed the notion of the brain as a reticulum, or a syncytium—a continuous network of interconnected cells and fibers.

Shortly thereafter, Spanish anatomist Ramón y Cajal, using the same Golgi method, determined that the brain was made up of individual cells, the neurons. He speculated that they were connected to one another through a synapse, which implied that individual units of the brain function separately but in an interconnected way. The term *synapse* was coined by a British physiologist, Sir Charles Sherrington, in 1896 following communication with Cajal. Over time, the synapse model prevailed, bolstered by the discovery of chemical messengers called neurotransmitters that provided the means of electrochemical information transfer from one cell to another. Observations of the importance of the synapse have led to increased understanding of the role of synaptic remodeling in learning, the capacity for synaptic regeneration and reorganization, and evidence that synaptic loss is an early manifestation of memory dysfunction as occurs in Alzheimer's disease. Mutations in the proteins that form the basis for synapse structure are known to be a cause of autism and schizophrenia.

Neurons

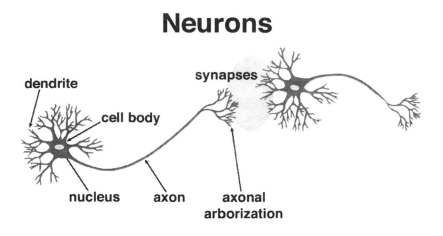

It isn't a question of who was right or wrong in comparing and contrasting the work of Golgi and Cajal. It really amounted to a difference in interpretation of their experimental observations. Both were outstanding scientists who shared the Nobel Prize for their work in 1906. But it was Cajal's anatomic interpretation and leap of imagination that framed the backdrop for most of the subsequent work on brain structure and function.

Trillions with a T

These observations bring together the fields of neurology and psychiatry and elucidate useful approaches to further understand these serious conditions. One can argue that the complexity of the brain results in a structure more marvelous than the universe. It is estimated that the brain contains a hundred billion nerve cells, each cell having up to ten thousand connections from other cells, resulting in ten trillion to as many as a hundred trillion synapses.

The synaptic regions where the connections are made from one nerve cell to another contain up to a thousand different proteins that are involved in the depolarization of the terminal, presynaptic release of a neurotransmitter that acts on the other side of the synapse—postsynaptic—each protein capable of modifying synaptic function in relation to mental activity and learning. Experimental studies show that synapses alter shape and strength after changes in firing pattern. Such modulation of synaptic strength occurs with changes in number and shape of synapses. This can occur over a few minutes or hours as in the process of laying down new memories. This has led to the concept of a constant changing, or plasticity, of brain structure and biochemistry as a result of experience. It is an indication that the brain is a wet, organic structure and not a computer.

This has led to the concept of a constant changing, or plasticity, of brain structure and biochemistry as a result of experience. It is an indication that the brain is a wet, organic structure and not a computer.

Excitatory Glutamatergic Synapse

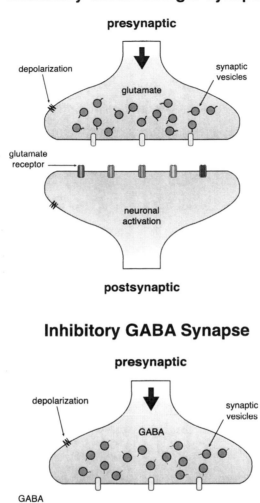

Inhibitory GABA Synapse

An understanding of the basic structure and function of synapses has led to the discovery of many drugs now used in psychiatry and neurology. Such drugs act to facilitate, inhibit, or modulate actions at the synapse. Glutamate is released at excitatory synapses to act upon several different receptors in the dendritic regions of post-synaptic connectivity. Its release is essential for the laying down of new memories.

GABA, an inhibitory neurotransmitter, has a role in the epileptic brain, with the result that many effective drugs increase the inhibitory bias. Other systems release the biogenic amines dopamine, norepinephrine, and serotonin, all important sites for drugs in treatment of Parkinson's disease, depression, obsessive-compulsive disorder, and post-traumatic stress disorder. As we shall see later, synaptic modulation also has a role in understanding disorders like Alzheimer's disease, Parkinson's disease, autism, and schizophrenia.

Psychoanalysis and structural diseases of the brain

A second example of dualism was the emergence of differences in perception of mental illness as a brain disorder. As is well known, Freud proposed near the turn of the twentieth century a psychoanalytic approach to understanding brain function and dysfunction. At the same time, in Germany, Emil Kraepelin and his student, Alois Alzheimer, proposed that structural changes might be found in examination of the brains of patients by autopsy and microscopic examination using methods Cajal and Golgi had developed. Freud and Kraepelin were both born in 1856, one in Austria, the other in Germany, locations central to the emerging basic sciences of physics, chemistry, astronomy, and astrophysics, as well as the medical sciences of biochemistry, morphology, embryology, and the like.

During the first decade of the twentieth century, Alzheimer followed a fifty-one-year-old woman for a period of four to five years; she had a dementing illness characterized by paranoia, memory loss, and language difficulties. Upon her death, brain examination using Golgi's silver stain demonstrated for the first time specific changes in the brain, consisting of neuronal (neuritic) scars called plaques and intracellular (neurofibrillary) tangles.

The plaques were found in the substance of the brain separate from nerve cells but often surrounded by degenerated debris from adjacent neurons, whereas the neurofibrillary tangles were found in the cytoplasm of the neurons, leading to obstruction of the normal function of the cells. Plaques and tangles became the diagnostic criteria for what Kraepelin in 1910 called Alzheimer's disease.

Once again, it was not a simple matter of whether the approach used by Freud or Kraepelin was right. Both contributed profound concepts that remain guideposts for how we interpret brain disorders and how we treat them.

What was Freud thinking during these years?

Ernest Jones, in *The Life and Work of Sigmund Freud*, writes that Freud stated, "I have no inclination at all to keep the domain of the psychological floating, as it were, in the air, without any organic foundation. But I have no knowledge, neither theoretically nor therapeutically, beyond that conviction, so I have to conduct myself as if I had only the psychological before me."[17]

In 1905, Freud is quoted as saying, "To avoid any misunderstanding I would add that I am not attempting to proclaim cells and fibers, or the systems of neurons that nowadays have taken their place, as psychical paths, although it should be possible to represent such paths by organic elements of the neurone system in ways that cannot yet be suggested."[18]

In 1917, Freud opines, "Psychoanalysis hopes to discover the common ground on which the coming together of bodily and mental disturbances will become intelligible. To do so it must keep free of any alien preconceptions of an anatomical, chemical or physiological auxiliary hypothesis."

Clearly, Freud was aware of the controversy.

Over the course of the twentieth century, the development of concepts of psychoanalysis spread broadly throughout the Western developed world, including Europe, Britain, and the United States. They were particularly prominent in Boston and New York up until the end of the Second World War. Nearly all of psychiatry practiced in the Northeastern United States was built upon Freudian psychoanalytic hypotheses.

In the meantime, further studies of the brain pathology in gross and microscopic detail identified a wide variety of different changes associated with mental dysfunction—other dementias occurring with neurosyphylis, vitamin deficiency,

---

17   Jones, E. *The Life and Work of Sigmund Freud.* Published 1953 by Basic Books, Inc.

18   Freud, Sigmund. *Collected Works of Sigmund Freud.* Published July 31, 2007 by BiblioBazaar.

metabolic disturbance, and in degenerative diseases like Parkinson's disease and frontotemporal dementia. It is not surprising that until recently, the fields of neurology and psychiatry were increasingly separated in terms of patient care and in the education of medical students and of specialists in each of these fields. The expansion of research and clinical activities in each discipline led to an increase in the number of separate departments, with minimal overlap in clinical, educational, and research activities.

The separate approaches of Freud and Kraepelin gave birth to these two competing approaches to brain study in the first half of the twentieth century.

A new convergence in neuropsychiatry

Conditions once considered "psychogenic," or driven by experience and adverse events, are now yielding to biological, biochemical, and genetic study. Two conditions, autism and schizophrenia, which in the past were examined principally by studies of psychological profiles and life experiences, are currently the focus of intense efforts to establish a biological basis. Both of these conditions are influenced by familial, presumably genetic, factors. Schizophrenia is common—up to 1% of the population in all parts of the world. Autism prevalence is rising, from 1 in 150 to 1 in 44 among the populations examined in developed countries, raising concerns about environmental and immunological factors. While vaccines were once blamed, this theory is now widely discounted.

Genetic studies in both conditions have revealed dozens of susceptibility or causative genes. The spectrum of autistic disorders is now known to encompass more than fifty primary genetic factors, by which is meant that a disordered, or mutated, gene function is associated with the clinical phenotype. Many of the genes associated with these conditions encode one of the synapse-associated proteins. There is particular interest in the role of developmental synaptic alterations, where it is hypothesized that excessive synaptic development occurs. For example, structural proteins that hold the synapse together, such as the neuroligins and neurexins, have been shown in mutated form to lead to the autism spectrum disorder [see figures]. Presumably, these genotypic failures lead to the autism phenotype via abnormal connectivity throughout wide portions of the brain.

Autism's Origin?

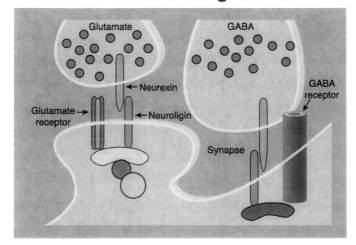

Critical Connection

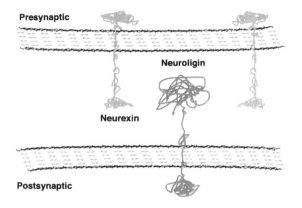

In recent years, a powerful technique has emerged called genome-wide association studies, or GWAS. Large numbers of subjects, up to or exceeding a hundred thousand symptomatic individuals, are compared with population controls to find gene variations that accompany disease manifestations. GWAS has been used effectively in autism, schizophrenia, and in disorders such as multiple sclerosis, epilepsy, and type 2 diabetes mellitus. These studies change the architecture of the phenotype-genotype landscape to the point that the burden of proof depends now on further deep genomic sequencing to define precise gene-protein anomalies. Most of the major psychiatric disorders are yielding information potentially relevant for new therapeutic approaches. An exception to this has been the great disappointment in locating these genetic influences in major depression and in bipolar disease, but these too will likely yield to new approaches.

Most of the major psychiatric disorders are yielding information potentially relevant for new therapeutic approaches. An exception to this has been the great disappointment in locating these genetic influences in major depression and in bipolar disease, but these too will likely yield to new approaches.

Dualism and the big question: Why are we here?

Perhaps the most profound example of dualism lies in the matter of how we perceive ourselves in relation to the world, our planetary system, and the universe.

Sean Carroll, a noted physicist from the California Institute of Technology, has just published a provocative, deeply penetrant, and analytic book entitled *The Big Picture*.[19] It addresses the continuing controversial and often adversarial relationship between secular humanists, who view the universe as a natural happening, and the religionists who view it—and us human beings—as caused by an external entity, who in most instances we refer to as God. Terms of intermediate significance arise from this perspective. The doctrine of theistic evolution holds that God did it but used the principles of natural selection and fitness to produce the variety of life forms, while believers in intelligent design argue that the complexity of it all is too overwhelming to have happened accidentally.

---

19   Carroll, Sean. *The Big Picture*. Published May 10, 2016, by Dutton.

"Faith is a gift," said one of my cynical colleagues. "You either have it or you don't!" Perhaps there is more truth here than at first might be implied. Of course, faith is a terrible curse as well when it breeds violence, slaughter, and perversion of individual human rights.

Carroll is the most convincing diplomat that I have come to know about in this arena. Richard Dawkins, Samuel Harris, Daniel Dennett, and Christopher Hitchens, sometimes referred to as the "four horsemen of the apocalypse," are well-known angry protagonists of the view that atheism is both a true and necessary belief system to bring order to our world.[20] Carroll, on the other hand, carefully weaves evidence for "poetic naturalism" that yields to awe—*how majestic it is*— and to wonder—*I'll try and find out how it works.*

The most challenging zone of conflict resides in perspectives about brain, mind, and spirit. It is difficult for those with a theist bent not to believe in the reality of an afterlife, after the brain is dead. Can there be a spirit of personhood that survives material demise? This again is not a matter of fact, study, or research, but rather an article of faith. Sadly, perhaps, there have never been provable incidents of consciousness beyond complete brain death, although many muster examples of near-death experience as proof. To the neuroscientist, these only represent lingering circuits of brain activity likely in the hippocampus or other memory-oriented brain sites.

As my classmate at EMC, Dr. Joe Longacre, notes, "When there is no longer any evidence of brain activity on an EEG,

---

20   (https://www.youtube.com/watch?v=cTjHf77FqTI)

we consider the patient dead, and can with good conscience withdraw life support. In other words, when the mind no longer works, and they are unable to think, they "are not", to paraphrase Descartes. Or in Latin, "non cogito, ergo non sum"."

In the end, I conclude that, both in fact and in truth, there can be no convergent notions that will satisfy both parties in the debate. What we can call for is mutual respect, realizing that there are two great mysteries that may never be resolved: how and what is the universe, and how is it that our brains are fashioned so that we can penetrate these deep questions, both who we are—consciousness—and what is it all about?

Whatever one's perspective, there is much to do to preserve our earth and to fight poverty, anarchy, and disease in order to leave our world a better place than we found it.

Perhaps that ought to be enough.

Reflections on Book I

In these essays, spanning nearly thirty-five years, I found myself returning to the controversies surrounding scientific discovery and its impact on religious beliefs. Of great concern to me is the deepening suspicion in our society regarding the scientific method in explaining the origins of human behaviors.

Many of our religious predecessors placed the earth at the center of God's great creation. The Genesis account documented Adam and Eve as the first humans, and believers declare that all humankind has descended from them. According to the lineage recorded in the Old Testament, one can trace subsequent generations and estimate that the human family is about six thousand years old and that the world was created in a literal period of seven days.

Consider the recent opening of Ark Encounter in Kentucky, funded by a Christian evangelical creationist who asserts that the biblical flood was an actual event. The centerpiece is a replica of Noah's ark, which is based on the dimensions provided in the Bible; the gigantic ark is seven stories high and one and half football fields long. Shockingly, photomontage images show dinosaurs roaming the countryside, presumably to later be destroyed by the flood.

Although discordance still exists concerning a number of concepts, science has successfully changed the way nearly all of us think about some previously held beliefs. Most believers no longer think that the world is flat, that the earth is the center of the solar system, or that our civilization is only a few thousand years old. The universe is fourteen billion years old.

Our earth is a speck at the periphery of a modest-sized galaxy we call the Milky Way. The fabric of our universe yields to post-Newtonian, post-Einsteinian concepts of atomic structure, quantum mechanics, and dark matter.

The concern for some is that scientific inquiry runs counter to the creationist, intelligent design conceptualization of the world and that scientific evidence risks corrupting religious faith. They believe that the precision of the universe's infrastructure is proof of a creator, God, and fear that concepts of Darwinian evolution will betray a commitment to religious belief.

For me, this brings to mind all those who believe in the science that has led to radio, TV, cell phones, electric cars, global positioning systems, radar, air-traffic control, and the substructure of our modern world, but who allow a literal interpretation of the Bible to get in the way of allowing science to teach us about intelligence, consciousness, sexuality, climate change, and evolutionary principles. Most troubling, perhaps, is that these topics are too often closed off from serious discussion and the actual evidence that supports these scientific claims goes unappreciated or denied relevance.

Addressing these issues has been a personal quest for more than sixty years, since I left home to attend the University of Alberta for premedical studies. Over the years, I have attempted to approach these dilemmas in the talks I've given.

Let me say now with conviction and assurance that science will always trump religion when it comes to experimental observations and the findings that emerge with careful attention to experimental hypotheses. In scientific study, the methods and results used to claim an assertion must be validated by others and are often soon replaced or amended by better scientific experiments.

A fundamental characteristic of consciousness and self-awareness is the curiosity to explore and hypothesize in a determined effort to understand and explain how it all came to be. It turns out that our brains are extraordinarily complex, and it may be surprising to realize that everything we know about the nature of things has come through the imaginations made possible by our brains.

The dogmatic efforts of religious establishments to overturn the discoveries of Galileo, Copernicus, Newton, Darwin, and Einstein have repeatedly been forced to yield to scientific inquiry. These are examples that illustrate the futility that arises from the use of ancient scriptural texts, which are then shown to be irrelevant in light of today's ideas and inventions.

"The great tragedy of science
is the slaying of a beautiful hypothesis by an ugly fact."
—Thomas Huxley (1825-1895)

Everything in the universe is made up of the same elementary particles. Living versus non-living depends fundamentally upon whether an entity can reproduce itself.

Does it matter what I believe? Given the uncertainty about so many things in our lives, how might what I believe affect what I do?

My response, as a dean, educator, and healthcare administrator, is to take a practical turn and inquire how such beliefs influence attitudes toward public policy, health care, and other civic duties. How should we take action to address poverty, healthcare disparities, chronic illness, and end-of-life care?

For those of Christian faith, I do believe that the Jesus model remains the best example to follow in our daily goings-about; it offers a set of principles that are eternal in application. Here the New Testament examples of Jesus' life and focus and of the emerging Christian Church still provide for me the best examples of how to live our lives.

Book II. Medicine as a Profession

CHAPTER 7

University of Alberta Centennial Celebration Talk

October 2008

I have an affinity for the University of Alberta, having earned my medical degree there in 1962. The province, established as part of the Dominion of Canada in 1905, moved presciently soon thereafter to establish a university in Edmonton in 1908.

In this speech marking the hundredth anniversary of the university, I mention having just returned from a sabbatical. After stepping down as dean of Harvard Medical School in July 2007, I spent a year in Boston writing my memoir and working with colleagues and students on the diseases of an aging population. I returned to the medical school as the Edward R. and Anne G. Lefler Professor of Neurobiology.

Then, Now, and When? Thoughts on the Future of Biomedical Science

Thank you for inviting me to join you today. It is a thrill to participate and to see all of you here. I have just returned from a wonderful sabbatical that gave me time and freedom to reflect on the current state of science and medicine and to ponder some of the challenges, problems, and opportunities that lie ahead. I want to share some of these thoughts with you.

A few years ago, shortly after the Berlin Wall came down and the Cold War appeared to be ending, Francis Fukuyama wrote a book entitled *The End of History and the Last Man*[21]. He suggested that we were entering a new phase of history, one that signaled the end of strife and the triumph of democracy. We all know how wrong that was.

Some have compared recent progress in biology to a similar endpoint, implying that we have reached a certain stage of maturation and knowledge, as if there might be an end to biology. Yet we know that for every discovery made new questions are raised.

I have entitled my talk "Then, Now, and When?"

I want to contrast the knowledge transmitted to me during my medical school experience at the University of Alberta in the years before my graduation in 1962 with current advances, and to send along the message of hope and challenge that lies before us. We have accomplished a great deal in biomedical research and therapeutic discoveries, but as I hope to convince

---

21   Fukuyama, F. *The End of History and the Last Man*. Published 1992 by Free Press, New York.

you, we remain uncertain about aspects of the progress we expect to see in the future.

With that in mind, let me recount four areas of current perplexity and opportunity in the life sciences that describe the current state of our understanding, and which offer special opportunities for research. I will include broad perspectives in the life sciences with examples ranging from fundamental science to health care. I will illustrate these points by focusing on four areas: genetics, infectious diseases, neuropsychiatry, and healthcare policy.

Genetics

Watson, Crick, and Wilkins reported the structure of DNA in 1953, arguably one of the most important discoveries ever in biomedical research. By 1962, when I graduated, we knew a great deal about Mendelian genetic inheritance and of diseases dominant, recessive, and X-linked that awaited analysis with emerging technologies. The techniques for RFLP analysis were published in 1980. RFLP stands for restriction fragment length polymorphism; in RFLP analysis, pieces of DNA are created to help in determining which gene or genes might be responsible for a particular disease. In 1983, the first linkage to chromosome 4 in Huntington's disease was discovered—in my group at the Massachusetts General Hospital.

During the 1980s and 1990s, the genes for many mono-genic disorders were identified, including those for Duchenne muscular dystrophy and cystic fibrosis. Yet even today, we have no general gene-based therapy treatment for any of these disorders.[22] Despite the progress made in these single-gene mutation disorders, we remain far from defining all the genes responsible for complex disorders like Alzheimer's disease, diabetes, schizophrenia, and autism.

Tip of the iceberg

You might recall that the "final" description of the human genome, published in 2003, was sometimes said to be the beginning of the end of human biology. That was five years ago. In the meantime, we have learned that the example of the genome as a final common pathway for the regulation of DNA-RNA protein regulation, as written in the triplet base code, is only the tip of an iceberg in terms of understanding the regulatory apparatus that controls our bodies and our minds.

In the meantime, we have learned that the example of the genome as a final common pathway for the regulation of DNA-RNA protein regulation, as written in the triplet base code, is only the tip of an iceberg in terms of understanding the regulatory apparatus that controls our bodies and our minds.

---

22   Thankfully, this statement proved wrong when in 2013 a biotechnology company, Vertex, was given approval for a drug that dramatically improved care for individuals with cystic fibrosis, although only in 5-7% of patients. Subsequent drug development has led to approval of additional therapies for cystic fibrosis. Since then, additional drugs have become available.

We now know that there are short interfering RNAs—siRNAs—whose discovery led to the awarding of the Nobel Prize two years ago to Craig Mello from the University of Massachusetts and to Andrew Fire at Stanford. Mello and Fire discovered gene silencing, a process that allows cells to selectively turn off specific genes. This discovery has enabled research focused on suppressing the expression of specific genes and has the potential to form the basis for new treatments.

In addition, investigators have uncovered an incredible new mechanism of gene regulation by a new set of influences, often contained in the "junk" DNA or in introns, which regulate many processes of protein biosynthesis. These are the microRNAs. MicroRNAs are endogenous 21-nucleotide RNAs that can pair to sites in the messenger RNA of protein-coding genes to suppress the expression of these messages. For the Harvard Medical School community, this year's announcement of the Lasker Award for fundamental research was particularly exciting. It was awarded to Gary Ruvkun for his initial discovery of microRNA in C. *elegans*, a round worm whose entire genetic makeup has been elucidated. Gary is in the Department of Genetics at Harvard Medical School and in the Department of Molecular Biology at the Massachusetts General Hospital.

We've also come to know that there are intron genes that encode microRNAs, and that epigenetic influences through cis and trans regulation have a major impact on the outcome of the genome's directives in the production of proteins and in their regulation. Last week in the journal *Nature*, a landmark collaboration between scientists at MIT and Harvard showed an entirely new pathway of integration of biological functions activated by the microRNA system. The investigators used quantitative mass spectrometry to measure the responses of thousands of proteins. The findings indicate that microRNAs act as rheostats to make fine adjustments to protein output. This work, carried out by a group led by David Bartel at the Whitehead Institute of MIT and Steve Gygi of the Department of Cell Biology at Harvard Medical School, was published simultaneously with a report from Germany in the same issue of *Nature*. The implication of these findings is truly astonishing, throwing open an entirely new way of thinking about regulation of the complexity of gene expression. This field, with other discoveries, is now encompassed by the term *epigenetics*.

Another example from the field of cancer research comes to mind.

From Philadelphia to Gleevec

Let me tell you the story of the Philadelphia chromosome. Just fifty short years ago, Peter Nowell, at the University of Pennsylvania, and David Hungerford, at the Fox Chase Center in Philadelphia, noticed a small additional fragment on chromosome 22, often associated with chronic myelogenous leukemia, or CML. In 1973, it became clear using new technology that this fragment was the result of translocation [when one piece of a chromosome breaks off and attaches to another or when pieces from two different chromosomes trade places]. The breakpoint of this translocation was shown by David Baltimore, who won the Nobel Prize for another discovery, and Owen Witte, both at MIT at the time. Witte went on to show that this mutated gene, the *abl* oncogene, produced an abnormally active form of tyrosine kinase, a finding that led to an inhibitor of tyrosine kinase and a successful treatment for CML, Gleevec.

An analysis of the human genome reveals at least five hundred tyrosine and serine kinases in total, each of which is a potential target for understanding cancer and for developing new therapies.

Genetics and genomics is truly an area where Pandora's Box has been thrown wide open.

Infectious diseases

In 1962, the polio epidemic had been largely controlled, thanks to technology developed by John Enders and Fred Robbins that allowed the virus to be cultivated in renal cell tissues. This paved the way for the development and introduction of polio vaccines by Jonas Salk and Alfred Sabin—who, by the way, could not stand each other. Today, we are within reach of eliminating polio from the face of the earth, were it not for political and cross-cultural issues that face investigators and physicians in Africa, Pakistan, and India. This is a whole other story related to public perceptions of vaccine administration.

Over the next forty years, vaccines were successfully developed for a host of bacterial and viral diseases including pertussis, mumps, measles, pneumococcal pneumonia, and for herpes virus to prevent cervical cancer—the Nobel Prize for 2008 announced just last week. Yet today, we face a major challenge that I will illustrate with HIV/AIDS.

The first patients with HIV were described just over twenty-five years ago. I can recall my first awareness of this disorder when it was discussed during a brain pathology session at the Massachusetts General Hospital in 1982. Scientists moved quickly to identify the full genome and many of the functional genes and regulators for HIV virus replication. It was thought at the time that it would be a short while before we would have a preventive treatment in the form of a vaccine. This proved to be extraordinarily difficult. Although drugs developed in the 1990s to treat HIV/AIDS have been quite successful, within the past few weeks we have witnessed a declaration on the part of Dr. Tony Fauci, director of the National Institute of Arthritis and Infectious Diseases, that funding for the work toward a vaccine is being suspended given the tragically difficult problems that have emerged in this research. An article regarding the controversy that appeared in *Science* a few weeks ago said, "In canceling the Partnership for AIDS Vaccine Evaluation (PAVE) study, which would have cost $63 million, Fauci challenged researchers to come up with a 'lean and mean' alternative. Fauci, who has lately faced intense pressure from AIDS vaccine investigators to put more money into fundamental research, says so much confusion exists about what a vaccine should contain that proceeding with this particular vaccine was simply too dicey. 'Given the fuzziness of all this, I'm just not willing to go ahead with such an expensive trial.'"

New ideas welcome

These facts point to the incredible challenges and difficulties that scientists experience in understanding infectious diseases. It teaches us about how difficult it is to work with mutating organisms and reminds us about the risk to our survival when challenges arise, whether in the form of agents such as the Ebola virus that originated in Africa or a pandemic flu virus of the avian flu variety from Southeast Asia. We now know that a similar avian virus was responsible for the influenza pandemic that killed millions of people around the world in 1917 and 1918. There are plenty of opportunities to apply new ideas to research on emerging infections that threaten the entire survival of the human species.

Let me tell you another story about today and tomorrow.

When I was in medical school, gastric and duodenal ulcers were linked to stress as part of the emerging field of psychosomatic medicine. Their treatment consisted of gastrectomy and vagotomy to remove the source of stomach acid. A surgeon, Ove Wangensteen of the University of Minnesota, developed gastric cooling to stop bleeding after ulceration.

Then came a couple of Australians, who the mainstream considered slightly out of touch, showing that gastric ulceration was associated with a new form of bacteria *Helicobacter pylori*. Suddenly, surgery, histamine receptor blockers, and proton-pump inhibitors yielded to antibiosis that removed the bacteria from the gut.

But that's not the end of the story. We now hear of an association between asthma and removal of *H pylori*. Furthermore, it's become clear that humans are the nesting ground for thousands of bacteria, whose identity is largely unknown and which have doubtless been with us for millennia—mostly for the good, we're told—whether in the gut or the mouth or indeed, on the skin.

Neuropsychiatry

In 1962, four classes of drugs emerged as effective in neuropsychiatry: lithium, whose mechanism of action remains largely unknown; chlorpromazine, discovered by accident; the first of the benzodiazepines—Valium; and the tricyclic antidepressant amitriptyline. In neurology, tremendous advances had been made in antiepileptic drugs. Dilantin was introduced in 1938, the year I was born.

Look for a moment at neuroscience today. We know now that there are at least a thousand different proteins in synapses of the mammalian brain, which form the critical communication between nerve cells. Imagine the requirements for the collaboration that will be needed to unravel this complexity.

The Power of Collaboration

Great advances have been made in discovering genes for monogenic disorders like Huntington's disease and neurologic disorders including Alzheimer's disease. The genes for these two disorders were discovered at Harvard Medical School by scientists working in our hospitals. Although we now understand more than a thousand disorders of this type, the complexity of brain organization and development has made finding the genes responsible for autism and schizophrenia difficult. Recent work from Chris A. Walsh's laboratory at Boston Children's Hospital, in collaboration with Mike Greenberg's group—Mike is now chair of the Department of Neurobiology at Harvard Medical School—points to the fact that multiple genes are associated with autism. Their work, published in *Science* just two months ago, illustrates the power of collaboration. The paper includes twenty-five authors from many different settings around the world. They included scientists from the Broad Institute of Harvard and MIT, which provided the data management and statistical analysis, and from the Boston Autism Consortium, which includes four Boston teaching hospitals, as well as colleagues working in Saudi Arabia, Kuwait, Turkey, and Pakistan.

I can't tell you how much it pleases me to see collaborative associations like this that bridge old rivalries and that cut across national and international boundaries. This is the future of much of the best clinical genetic work.

I can't tell you how much it pleases me to see collaborative associations like this that bridge old rivalries and that cut across national and international boundaries. This is the future of much of the best clinical genetic work.

Meanwhile, as I know many of you are aware from the enormous publicity surrounding the apparent increased frequency of autism, some parents question the role of childhood immunization. Here we are faced with a public health issue in which many respectable people believe that childhood vaccination may be associated with increased incidence of autism. As recently as a few weeks ago, we had an outbreak of measles in children who were not protected. These are public health issues that cut across many avenues of disagreement. Part of our challenge is going to be how to work in an information overload society where information is instant, often uncensored, and frequently wrong. How do we deal with this?

Health Care Policy

Medicare was first introduced in a province-wide experiment in Saskatchewan in 1962 under Tommy Douglas. The doctors went on strike for twenty-three days fearing intrusion of the government into their practices. Today, a single-payer system is accepted across Canada with joint agreements between the federal and provincial governments.

How is it in the U.S. today?

Healthcare reform failed under Clinton in 1994 and did not reach national consideration until this year. The issues are well publicized and can be summarized in nine points:

- The United States spends 16% of the gross domestic product on health care.

- This amounts to approximately $8,000 per person per year, nearly twice that of Canada, the U.K., and Japan.

- There are presently 45 million people without insurance, another 30 to 40 million who are uninsured for part of each year through lapses in coverage between jobs or during periods of growing unemployment, and many more whose insurance is inadequate to meet the expenses of a major illness.

+ Estimates indicate that up to 50% of personal bank-ruptcies are triggered by excessive healthcare costs or catastrophic illnesses.

+ Many leading institutions, including major university hospitals, have grown enormously in size in clinical care and research but have become unable to provide equal growing capacity for primary care services.

+ An aging population with anticipated needs in manag-ing of chronic disease often have no "medical home" to coordinate care, which results in fragmented, episodic care given mostly by specialists who communicate little or not at all with each other.

+ A new model for minor medical care is developing in pharmacies and in stores like Walmart, with ineffective linkage to physicians for follow-up treatment. This business model is generally challenged by the deliverers of status quo medicine without adequate exploration of the possibilities this model represents.

+ Fewer medical school graduates are going into primary care.

+ The dearth of primary care doctors has resulted in a proposed solution to increase the physician workforce by admitting more students and has been implemented by many medical schools, based on a directive issued by the American Association of Medical Colleges.

You know what? All the members of Congress are fully covered, and they have great pension benefits. Why do they not feel a responsibility to their constituents? What if we had a national lottery and took away health insurance from 20% of the members of Congress? We'd get some action then. It is a disgrace that a wealthy country like the U.S. has not had sufficient social conscience to fix this problem.

Fortunately, we have had great collaboration in Massachusetts involving hospitals, insurers, business, and government, and we are deep into an experiment on a new universal health plan that shows great promise. It was hard to put together, but a coalition representing all relevant parties got it done.

Even with universal coverage, we have a serious delivery problem. We do not have the primary care infrastructure in Massachusetts or anywhere across the United States to deliver the care our citizens deserve. Under urgent consideration today is the question of whether we have enough doctors and dentists to care for our patients, particularly if the country moves toward a new scheme of universal health coverage.

A Final Piece of Advice

Let me finish with a practical note. I would like to encourage those of you undertaking research careers in the basic biomedical sciences to find time once or so each month to read the latest issue of *The New England Journal of Medicine* to savor the world of clinical medicine and the impacts it can have on your work.

For those heading toward clinical careers in medicine and dentistry, pull out a copy of *Nature* or *Science* or *Cell* from time to time and read the introductions to the papers relevant to biomedicine. You will be astonished at the insights that will come your way.

Thanks for allowing me to share these thoughts with you.

Postscript

Reflecting on the commentary of my speech just nine years ago, I marvel at the progress that has been made in so many areas of biomedical research. A few examples will illustrate these advances.

Beginning with neuropsychiatry, the genes for many Mendelian inherited forms of disease have been discovered in Alzheimer's disease, amyotrophic lateral sclerosis (Lou Gehrig's disease), autism, and recently in schizophrenia. As one examines the great opportunities of the twenty-first century, I have become enamored of the notion that this will be known as the *Century of the Synapse*. Whether in the scrambling of sensory inputs as occurs in synesthesia, the failure of synaptic pruning in autism and, very recently uncovered, evidence for excessive pruning of the developing brain in schizophrenia, there emerges strong evidence for focus on this fundamental aspect of brain function.

There are promising new leads in gene therapy with the discovery of increasingly potent and selective adenoviral vectors for transport of proteins to correct protein disorders in the central nervous system. Almost certainly there will be some heroic discoveries in getting corrective proteins to sites critical for treatment of such disorders as amyotrophic lateral sclerosis and spinal muscular atrophy (SMA). Considerable commercial interest from companies like Biogen has made progress in this area very promising.

Perhaps the most astonishing recent discovery is identification of a technology for gene editing. The technique, called CRISPR/Cas9 (clustered regularly interspaced short palindromic repeats) allows the excision of DNA fragments and substitution of edited pieces of DNA. This raises issues including correction of mutated or faulty genes or the substitution of DNA loci for enhancing functions—increasing height, preventing baldness, enhancing cognitive capacity, and so on—each of which is loaded with biomedical ethical issues.

The critical importance of developing effective scientific and epidemiological studies that will generate ways to contain emerging viral threats cannot be overemphasized. We are familiar with avian viral plagues that wipe out ducks, geese, and chickens and have the potential to spread to humans. Similarly, we learned about the threat caused by severe acute respiratory syndrome (SARS), more recently Middle East respiratory syndrome (MERS), and in the past two years, Ebola. In the last few months, we have been made aware of the role the Zika virus plays in microcephaly and other birth defects. The rapid transcontinental spread of these vectors of disease is worrisome, particularly with global warming, where infectious disease transmitting agents—think mosquitoes— will flourish.

A remaining frontier of continued disappointment has been the inability to develop an effective vaccine for HIV. The issue has been addressed by the most fertile minds of infectious disease specialists, but the remarkable capability of the virus to mutate and transform has eluded the best ideas.

In healthcare, we find ourselves after an election year where the predominant Republican position is to destroy Obamacare and with it any of the many advantages that have been anticipated. The costs of healthcare continue to escalate, and no countering scheme has been offered. In my own state of Massachusetts, the efficacy of our healthcare system has been almost magical, with over 98% coverage and an interactive, effective, continuing collaborative effort between health care insurers—mostly still not-for-profit, health care providers, and government agencies with support from the business community. But here, as everywhere, the escalation in the cost of insurance is an emerging concern, with ever-rising copays for medical care and prescription drugs. Americans will be monitoring the changes sure to come under the new administration.

CHAPTER 8

Commencement, McGill University Health Science Convocation

May 31, 1994 [23]

In addition to medicine, the McGill Faculty of Medicine includes three allied health sciences schools: Nursing, Physical and Occupational Therapy, and Communication Sciences and Disorders. I was invited to speak at the graduation ceremony for all of these schools.

In the spring of 1994, I was approaching the one-year anniversary of my appointment as chancellor of the University of California, San Francisco. I was happy, as I always am, to return to Montreal and to McGill, the university that gave me my first real job in 1970. I established my neuroscience research career at McGill and gained recognition as I progressed up the academic ladder. My family and I have good memories of the years we spent in Montreal. Leaving for Boston and Harvard Medical School in 1978 was bittersweet.

---

23   This speech was published in *Canadian Speeches: Issues of the Day.* July 1994, Volume 8, Issue 4, p. 26.

Health Care Confronted by Anti-Intellectualism

Thirty-two years have passed since I graduated from medical school, and a lot has changed. But I am not old enough—although certainly gray enough—to talk to you from some elevated perspective of a life gone by. I'm still in the thick of a career in medicine and just as uncertain in many ways of what the future holds, as I suspect many of you are. I am reminded in addressing a graduating class of the comments attributed to Nathan Pusey when he was president of Harvard University. When asked why Harvard is such a fount of knowledge, he replied, "because the freshmen bring in so much and the seniors take out so little." I know that this is not true of McGill graduates.

I thought it appropriate to divide my comments into two parts. First, I will reflect on some of the current issues that we in the health professions face, whether defined in terms of an academic career, which I hope many of you will pursue, or in terms of the practice of the art and science of medicine, dentistry, nursing, or physical therapy, which the majority of you will undertake. Then, I want to give you some personal advice derived from my own experiences, which you may choose to heed or ignore, about how to make your lives as full as they can be, both professionally and personally. Finally, I promise to be brief.

A glass half full

Please forgive me if I sound like somewhat of an idealist, but I can only confess at the outset to a lifelong predilection to optimism and cheerfulness. This is no doubt accentuated by my belief that medicine and other careers in the health professions remain the best and most noble of all professions. These "weaknesses" have sustained me throughout difficult times and provided me with an enduring confidence that the future holds even more promise than the past.

It has been said that "there are no atheists in medicine; everyone thinks that he or she is God." Unfortunately, the profession as a whole has had some withering of self-esteem in recent years. Our patients ask more questions and insist on better explanations of the actions and side effects of the drugs they are prescribed. There are more constraints on the choices that can be made in delineating a career—choosing between primary care and specialty practice, for example. At least in the U.S., there is always the concern about making mistakes in the practice of medicine. Yet even with these constraints, your careers promise much.

It has been said that "there are no atheists in medicine; everyone thinks that he or she is God." Unfortunately, the profession as a whole has had some withering of self-esteem in recent years. Our patients ask more questions and insist on better explanations of the actions and side effects of the drugs they are prescribed.

Many have described the second half of the twentieth century as the biological revolution. The sixteenth and seventeenth centuries witnessed the renaissance in painting and sculpture, the eighteenth century the emergence of the genius of music, the nineteenth century the industrial revolution, and the first half of the twentieth century, the era of physics and chemistry. Now as we close out the twentieth century, we can only look back in awe at the accomplishments in cell biology, in genetics, and in the clinical advances in new treatments. This is the century that will be remembered for the atom, the computer, and the gene.

It is ironic, in a way, to consider that fifty years ago, toward the close of the first half of this century, physicists were perfecting the means for us to destroy ourselves with the atomic bomb. Now, toward the close of the second half of the century, we are on the brink of being able to change human-kind. Genetic engineering will, in your lifetime, have as great a potential for good or evil as atomic energy did in mine. As physicians, we need to stand guard over these great advances and be disciplined in our attention to the Hippocratic Oath: "to do no harm."

Advances and contradictions

The contradictions of our society are quite profound. On the one hand, we are witnessing the most remarkable emergence of new biology—the era of molecules, gene therapy, rational drug design, and new understandings of the genetic nature of our makeup. We are beginning to unravel the very nature of the brain and mind. On the other hand, we find a society removed from, and in large part ignorant of or indifferent to, these implications. Read the daily newspapers, survey the best seller lists; ours is a society infatuated with the new age, with anti-intellectualism, with books like *The Bridges of Madison County*.

The top of the fiction category on *The New York Times* best seller list for the past many weeks is *The Celestine Prophecy*, which claims to have discovered "the nine key insights into life itself," insights that include the observation that humans who increase their spiritual energy can become invisible and capable of passing into heaven at will. The best seller in the nonfiction category is a book based on a near-death experience, entitled *Embraced by the Light*. Imagine this: A literate, Pulitzer Prize–winning Harvard psychiatrist has written about the abuse of patients by aliens—not in the spirit of understanding modern applications of psychoanalysis, which might be of interest, but because he believes the stories are true.[24] We face a remarkable degree of rejection of the rational, the scientific. There exists an enormous dichotomy between the greatest era of biology and the growing dissatisfaction on the part of many that little of what science and medicine are about is relevant to societal or personal needs.

There is some basis for this skepticism. For example, despite the extraordinary advances in discovering the molecular structure of the HIV virus, we are still far from an effective treatment or vaccine. We believe that all our genes will be identified in the next fifteen to twenty years, yet we struggle

---

24   Mack, John E. *Abduction: Human Encounters with Aliens*. Published April 20, 1994 by Scribner. In it, he says he treated patients who claimed they had been abducted by aliens and insisted that it was a form of cosmic correction for our earth-polluting ways. Dr. Mack died in 2004 following an accident in London.

to enact laws in the United States to outlaw assault weapons, in the name of freedom for the few who do sport with them. The skepticism runs deep that we have not fulfilled our promises to conquer disease. We have advanced only a marginal amount in reducing the seriousness of the major cancers—breast, lung, prostate. The most common cause of lung cancer, smoking, is still an epidemic among our young people, and all of the available evidence about the adverse results of smoking seems to have had little impact on smoking behavior, particularly among young women.

There has also been a remarkable emergence of alternative medicine and of nonscientific forms of treatment, ranging from chiropractic to homeopathic.

Negative feedback

Then there is the perception of medicine as a treatment form gone amok. We are accused of keeping people alive when they ought to be permitted to die; we fail to recognize the extreme beauty of life and to permit a gracious exit. Jack Kevorkian becomes a hero. We recognize that rising costs of health care cannot be permitted to go on unchecked but fear the need to ration investigations and treatments.

The dissatisfaction continues within the profession of medicine itself. An increasing percentage of physicians express concern about their choice of the profession, and in recent surveys, as many as 40% have indicated they would have chosen other careers. Yet, in the U.S. this year, there were more applicants to medical schools than ever before, and many of the best and brightest who formerly would have chosen careers in business and law are now entering medicine.

The public is cynical of our successes and demands a prescription for immortality. We are encouraged to conduct more research to solve the degenerative diseases of our cells and tissues that accompany aging. Indeed, aging itself has become a disease. Unfortunately, skepticism about our science has also been fostered by unfortunate instances of scientific fraud and misconduct, an instance of which has penetrated deeply into your own community here in Montreal.[25]

Why this litany? Does the future seem unusually complex or bleak? How do we face it and become activists in bringing the best to reality? Proust said that the only thing that never changes is that every generation thinks that it has changed the most.

Ours has often been a misunderstood profession. We are present as the spectrum of life begins and we watch it end. We experience the joy of a new therapeutic accomplishment, and we witness the failure to change the course of fate. In the practice of our profession, we are the guardians of the soul and of the mind.

A few tips

Let me suggest a few things that you might consider as you accept the rewards and the challenges of this extraordinary role.

---

25 The reference to fraud concerns Dr. Roger Poisson of the University of Montreal, who falsified data in a study of breast cancer in women.

First, find a way to protect and nourish your intellectual vigor. In other words, do not graduate. Remain a scholar of medicine—to do otherwise will only guarantee obsolescence. Alan Gregg said, "A good education should leave much to be desired."

Perhaps some accomplishment will be found in a field that will complement your profession, as a musician or a painter. It is too easy to drift into complacency and to become the recipient, but fail to be a contributor to a form of creativity. Choose an area of medicine—or nursing, or dentistry, or physical therapy—in which you can become an expert, even if it is only so that you know more about the topic than anyone else. Everyone needs an arena within which to exercise intellectual experimentation.

Second, pay attention to your families, both to those who helped you accomplish what you celebrate today and those you have, or will have, brought into the world. Figuratively speaking, try to be home for dinner each evening. Even if you miss the joys and sorrows that are a part of every day in the life of those you care about and for, these emotions need to be experienced hands on, not related by second-hand telling.

Third, accept the increased responsibility dictated by the health profession's more vulnerable role in modern society to share openly and candidly with your patients what they can expect of you. Medical care has a large piece of prevention in it. Try to foster the idea of health and well-being, in addition to treating the sick and disabled. As part of your own contract with life, be sure to take care of yourself—adequate sleep, exercise, and time to relax and think.

> Try and foster the idea of health and well-being,
> in addition to treating the sick and disabled. As
> part of your own contract with life, be sure to
> take care of yourself—adequate sleep, exercise,
> and time to relax and think.

There is one interesting aspect to life that suddenly blossoms into reality at about my age—that one never really ceases to view oneself differently, regardless of age. And that who and what you are will change little over a lifetime. At your fifteen- and thirty-year reunions, each of you will be much as you are today—a thought that can both terrify and reassure.

> At your fifteen- and thirty-year reunions,
> each of you will be much as you are today—a
> thought that can both terrify and reassure.

Fourth, engage yourself fully in the responsibilities of citizenship in your community, in your country, and in the world. Our responsibilities extend to making the world a better place, not just by promoting health and wellness, but also by taking stands on the issues of our times and devoting time and energy to important causes. We will be increasingly challenged to broaden our horizons in the health sciences. Over-population; destruction of the environment; the threat of new chemical, viral, or bacterial plagues; and our survival as a species may be the most important problems of the twenty-first century. Physicians remain the most respected members of our society; we can use that reputation to advance the good.

Finally, maintain a sense of humor, an appreciation of the ridiculous. Laugh at yourself from time to time; it ought to be fun, and if it isn't, find out why and change it.

Godspeed to you all.

Postscript

One of the predictions that I made in this speech was thank-fully off the mark. It involves treatment for HIV/AIDS. I said, "For example, despite the extraordinary advances in discovering the molecular structure of the HIV virus, we are still far from an effective treatment or vaccine." In 1996, two years after I gave this speech, the combination treatment known as highly active anti-retroviral therapy (HAART) became available, dramatically changing the prognosis for HIV/AIDS patients. Research continues on developing an effective vaccine.

Sadly, there exists in the United States today a huge sus-picion among the public regarding the scientific approach to understanding. However, it can be ameliorated through good education and good works, driven by the recognition of the need to act.

CHAPTER 9

Bend the Curve

Warren Alpert Medical School of Brown University
Commencement Address
May 27, 2012

I am so pleased to join you today. I am deeply honored by this opportunity to share some thoughts with you. Congratulations to each of you on this special day.

I know that commencement speeches should be inspiring, optimistic, and most importantly, brief. I also know that I represent one of the last obstacles between you and your diploma.

Let's begin with inspiring. In polls asking about respect for professions and other jobs, doctors remain near the top of the list, along with nurses, of the most admired of professions.

I know of no other field of endeavor that offers so much variety and so many exceptional opportunities as the one you are embarking on today. At my age, I am perhaps a bit nostalgic, but I remain convinced that medicine, serving those we are privileged to care for and care about, is the noblest of professions.

My own graduation was exactly fifty years ago, and I cannot imagine choosing a more satisfying direction to have taken. Today, I don't feel much different than I did then, until I greet the guy in the mirror each morning and when I hear my grandson Cole ask, "Grandpa, why do you have all those wrinkles around your eyes?" That is a sign of wisdom, I reply, hoping that at age nine he will believe it.

My how time has flown by! But I refuse to think that we are separated by fifty years. Indeed, there is no gap separating us. We are part of a continuum, dating back over 2,500 years, of commitment to and willingness to adhere to the principles of the Hippocratic Oath. Medicine is a curve of progress, of achievements and of victories over disease and disability.

Bend with optimism

Events in life are commonly referred to as peaks and valleys, ups and downs, and so on, but rarely as curves. But I like this metaphor better. Life in your field of medicine is a series of curves, a change in direction more often gradual than abrupt, but each sufficient to alter forever the trajectory of a life-time. Being optimistic is critical to taking advantage of each bend in your curve. Let me briefly tell you about some of my experiences.

I was born in Alberta, Canada, raised on a dairy farm, with childhood aspirations to be a missionary doctor in China, India, or Africa. In medical school, my academic standing encouraged members of the school's faculty to support me in seeking a career in academic medicine, with a specialty in neurology. During that period of my life, I developed a great interest in research. This led me, after five years of clinical training, to re-enter student life at the University of Rochester, where I studied neuroscience.

I can trace my research interests to a particular group of patients I saw. Each of them had postural hypotension—fainting when they stood up quickly—a condition called orthostatic hypotension. Often idiopathic, and now sometimes called the Shy-Drager syndrome, or multiple system atrophy (MSA), it is a degenerative brain disease. These patients actually triggered my interest in research. I wondered why their blood pressure tanked when they stood up; they failed to show a compensatory tachycardia. Their autonomic nervous system reflex from baroreceptors to medulla back to the heart was not working. I wanted to find out why. This curiosity led me to study hypothalamic regulation of the endocrine system.

Taking the curves in stride

In my experiences, each event has been a curve, from boyhood on a farm, to pre-med, to medical school, to neurology, to graduate work, and finally, after fifteen years, my first real job—something my mother began to doubt would ever occur—with an appointment at McGill University.

Each bend in the curve came about through serendipity, guided by prescient mentors and the heroes I met who framed the context and reality of my ambitions. Life is a series of accidents, not plans, of unexpected moments never dreamed of.

I have entitled my talk "Bend the Curve." But, just a minute, you might be saying. I thought that was an economic term applied recently to the unsustainable costs of health care. That's right. If you Google the phrase, that's what comes up, and I want to describe how this is relevant to our consideration here today.

It is common to see a graph of healthcare costs over time showing a larger cost per individual and an increased percentage of the country's gross domestic product (or total output) being spent on the healthcare system and our patients. Now, the latter is not bad if it provides jobs and better health for our citizens, but when the money disappears into the health-care *system*—the inefficient, fragmented, specialty driven, corporate-profit-motivated medical-industrial complex—we should worry. Here the object should be to bend the cost curve downward by improving healthcare access and containing costs that are not conducive to better health.

Following directions

As I gaze out at your class, the graduates of the class of 2012, I think about how you are now embarking on a direction framed by that great event in mid-March called Match Day. That day has determined the bend in your curve and increased the likelihood of you getting to where you eventually want to go and, in every case, will allow you to pursue the directions that you've determined to be the best for you.

I have analyzed the career choices framed by your selections for residency training and find them representative of

U.S graduates in general. The largest number is going toward internal medicine, a smaller group in neurology and psychiatry, pediatrics, and OB/GYN, with a vast sprinkling of surgical subspecialties, radiology, and—I'm pleased to note—several of you will be entering family medicine and primary care.

So far, much of the trajectory of your curve has been more passive perhaps than active, more driven by the rules demanded to receive a professional degree, preparing you for the differentiation your life will unfurl.

Allow me to suggest some ways to bend your curve going forward to be prepared for the startling and unanticipated directions your careers will take.

First, enter the next stages of your experience with a deep commitment to lifelong learning. I was speaking with a colleague at Harvard Medical School about what to encourage young physicians to do. He suggested keeping your eyes open for the unusual, the unexpected. Every patient you see has a new story to tell; each will inspire observation, and each should be a teaching lesson. Look for the surprise in the story, for it is in this way that new discoveries are made. The best research remains that which is patient focused.

Second, you are entering a transformational moment in the delivery of health care in the U.S. Let me remind you of a few facts:

+ The United States spends 17% of the gross domestic product on health care;

+ This amounts to over $8,000 per person per year, nearly twice that of Canada, the U.K., and Japan.

- There are presently over 40 million people without insurance, another 30 to 40 million who are uninsured for part of each year through lapses in coverage between jobs or during periods of growing unemployment, and many more whose insurance is inadequate to meet the expenses of a major illness.

- Estimates indicate that at least 50% of personal bankruptcies are triggered by excessive healthcare costs or catastrophic illnesses.

- Many leading institutions, including major university hospitals, have grown enormously in size in clinical care and research, but have become unable to provide equal growing capacity for primary care services.

- An aging population, with anticipated needs in managing of chronic disease, often has no "medical home" to coordinate care, which results in fragmented, episodic care given mostly by specialists who communicate little or not at all with each other.

- A new model for minor medical care is developing in pharmacies and in stores like Walmart, with ineffective linkage to physicians for follow-up treatment. This business model is generally challenged by the deliverers of status quo medicine without adequate exploration of the possibilities this model represents.

- Not surprisingly, perhaps, fewer medical school graduates are going into primary care.

Joseph B Martin

To summarize my second point, spiraling costs, the general failure of a primary healthcare delivery system, over specialization, and a lack of access to the best care for the uninsured and those living in under-served, often rural, communities demand new approaches. You will be in the mainstream of this radical change. Many among us await with trepidation the outcome of the Supreme Court ruling regarding the constitutional validity of the Health Care Reform Act.[26] Whether or not the high court lets stand the current law, you will be in the midst of a new turmoil not seen in this country since the civil rights movement and the introduction of Medicare in the 1960s.

To summarize my second point, spiraling costs, the general failure of a primary healthcare delivery system, over specialization, and lack of access to the best care for the uninsured and those living in under-served, often rural, communities demand new approaches. You will be in the mainstream of this radical change.

Observe these events closely, but more importantly, I want you to participate in them. Grab hold and be part of a revolution that is now inevitable. Choose your position and fight for it.

Third, inherent in this evolving world will be new requirements for teamwork and sharing of professional opportunities

---

26   In June 2012, in a 5-4 ruling, the Supreme Court upheld the Affordable Care Act.

with others in the healthcare work force. Those of you in primary care and in specialties will need partnership with nurse practitioners and physician assistants, as well as pharmacists and social workers. You will observe the expansion of primary care into settings once deemed inappropriate—schools, perhaps pharmacies, and retail establishments. How can we team with nurses in schools to provide preventive care, immunizations, and minor health care?

Fourth, be a part of your communities beyond your own workplace. Join friends and neighbors in the activities of schools, churches, synagogues, and mosques. Let your children see that you care about the community you reside in and help them learn to exercise their gifts and skills as you do, to the betterment of the world.

Let your children see that you care about the community you reside in and help them learn to exercise their gifts and skills as you do, to the betterment of the world.

Fifth, you will experience a transformation in the imprint that science will bring to your practices and, in the case of those of you headed for academic careers, in your approach to science. You know of the power of genomics, still largely theoretical, but inevitably going to impact the personalized medicine and pharmacogenomics that will increasingly invade all aspects of good care. What drug will be most beneficial and safest for Mrs. Jones but dangerous for Mr. Smith? How will the inexpensive availability of your patient's fully characterized genome sequence lead to new informed consent rules, new biomedical ethical issues of who should know what?

Sixth, you are entering a world of globalized medicine. Great advances are being made in research and treatments in countries off the beaten path only a short time ago—India, China, Poland, Brazil, and South Africa. The patients you will see may well have traveled recently to these and other countries, as tourists or indeed as patients. We no longer hold the only source of sophisticated care. As travel has brought us closer together, it has amplified the possibility of widespread contagious events. This happened with SARS a few years ago, which spread quickly from Hong Kong to Vancouver and Toronto, and which remains a possibility with avian flu. I have no doubt you will witness the first truly effective vaccine for malaria, and will hopefully note improved access to treatments available for HIV/AIDS in every country. Be sure to ask your patients where they have traveled in the last six months.

Lastly, your generation will see the average lifespan approach eighty-five to ninety years. Accompanying the great success we have seen in treatment of hypertension, heart disease, arthritis, and the replacement of worn out hips, hearts, and kidneys, now we are on the verge of repairing brain damage. Here at Brown, you have heard of the remarkable work linking thinking to moving in paralyzed patients. Electrodes placed on the motor cortex with impulses recorded there are transmitted to computer-directed movements of the paralyzed limb. One can only imagine the direction this kind of work will take.

I hope, even in my lifetime, to see effective disease-delaying treatments for Alzheimer's disease and Parkinson's disease.

I have mixed emotions looking out upon you today. I have been witness to so much progress in the past fifty years, but the best is yet to come, and you will be there to see it. Go ahead and bend the curve.

Congratulations and Godspeed. And, finally a word from your alumni association—keep in touch.

CHAPTER 10

University of Rochester School of Medicine and Dentistry Commencement Address

May 16, 2008

To return to the U of R on this auspicious occasion was a special treat. The university served as my scientific beachhead when I studied there from 1967-1970, leading to a PhD in anatomy (neuroscience) awarded in 1971.

In May of 2008, the U.S. was six months away from the presidential election that would result in Democrat Barack Obama being elected to his first four-year term. Healthcare reform was being hotly debated between candidate Obama and the presumptive Republican candidate, Senator John McCain. I make reference to this and to an experiment taking place in Massachusetts requiring everyone in the state to obtain health insurance.

Under legislation passed in 2006, Massachusetts residents now have near-universal coverage. Some aspects of the Affordable Care Act, which was enacted in 2010 and has expanded health insurance coverage across the U.S., were borrowed from the Massachusetts plan.[27]

---

27   Portions of this speech appeared in the Boston Globe as an Op-Ed, May 27, 2008

Where Have All the Doctors Gone?

Where do all the doctors go? I'd like to suggest that there are three ways to examine this question.

First, where do today's graduating doctors go? Second, are we facing a shortage of doctors, and if so, in what areas of patient care? And third, where should doctors go or be seen in an election year with healthcare reform atop the national political agenda, along with the economy?

Where do today's graduates go?

I begin with a recent visit from a Harvard College graduating senior named John. He was completing premedical studies with an excellent academic record and a double major in economics and health policy. The purpose of his visit was to ask my advice about a career in medicine, as compared to entering the seminary for training for the ministry. He clearly was interested in doing good in the career he chose. Perhaps he had heard that at one point in my own career, I had taken a one-year leave of absence from medical school to undertake religious studies.

My thoughts focused rapidly on the extraordinary opportunities that a career in medicine offers. What other career offers so many opportunities?

You can choose to spend time in medicine or pediatrics or in surgery or neurosurgery. You can choose to work with pipettes, oscilloscopes, a PCR machine, or move into whole-genome mapping of your favorite disease. You can choose to live and work in Rochester, or Oswego, or Genesee, or Costa

Rica, or Rwanda. You might choose to work on MRSA or a vaccine for malaria. You might choose to work on a Native American reservation or open an office on Fifth Avenue in New York. Or you might get an MBA and start a new Internet or biotech company.

You can choose to work an eight-hour day as a dermatologist or anesthesiologist or settle for a sixteen-hour day in primary care. You can choose to take care of live people or look after dead ones—the field of pathology has grown in interest among medical students.

No other field offers so much. So where *do* all the doctors go? The current view is that many of you are hitting the ROAD—radiology, ophthalmology, anesthesiology, and dermatology.

I suspect that the graduates of U of R, like Harvard, show the most interest in internal medicine, pediatrics, and primary care. Most will specialize, however, and only a few will go into rural areas. Students from a minority background are more likely to return to their communities, often to serve the underprivileged.

Is there really a shortage?

Under urgent consideration today is the question of whether we have enough doctors to care for our patients, particularly if we move toward a new scheme for universal health coverage.

Interestingly, fifteen to twenty years ago, there were concerns about too many doctors, particularly in some specialties. Now, the Association of American Medical Colleges is requesting that medical schools increase enrollments by 30% over the next seven to ten years. There are fields like general surgery, where serious shortages are expected, particularly in smaller urban centers and rural districts. With an aging population, there will be an increasing demand for geriatric medicine.[28]

I was interested in an article last week in *The Wall Street Journal* about the expected shortage of neuro-ophthalmologists. There is grave concern about the lack of primary care doctors to work in settings where the patient load is high but the pay is low.

Every year, U.S. medical schools graduate about 16,000 students. We welcome another 6,500 foreign medical graduates into first-year residency slots. About 80% of these graduates will remain in the U.S., unfortunately often depriving their home countries of the workforce required to deliver adequate medical care there. Like their American counterparts, the majority of these foreign medical graduates also specialize. Fortunately, many of them seem willing to work in rural and underserved parts of the country.

What will we do to deliver the quality of care expected for and deserved by our patients? How will we increasingly focus on the importance of prevention and public health measures, encouraging parents to vaccinate their children, supporting major initiatives to stop smoking, and developing regimens for weight control that actually work?

---

28   By 2015, medical school enrollment had increased by more than 20%, made possible in part by the opening of more than ten new schools.

I strongly believe that the answer is not to train too many more doctors, but to give those we train the right jobs with pay that is commensurate with the contributions made. Perhaps we need to address the disparity in reimbursement, where doing procedures pays well but thinking deeply about a patient's problems has financial limitations. I am convinced that the new requirements in medical care will demand new models of healthcare delivery—a new focus on teamwork, where, for example, doctors, nurses, pharmacists, and social workers form efficient groupings to consider patient-centered care.

What's the role of doctors in an election year?

What is our civic responsibility, yours and mine, in an election year featuring a major reorientation to how we should deliver the best health care possible? This care needs to include greater access, particularly for the poor and uninsured, and at an affordable cost, which is in the country's best economic interest.

I like to refer to the five A's of healthcare, which need to be respected and considered in the effort to establish a more equitable and effective healthcare system: Access to Affordable, Affable, Accurate Advice.

What is our responsibility to work with our governments, our cities, and our health departments to accomplish this? Currently, it is estimated that forty-seven million Americans are uninsured. In Massachusetts, where an experiment is under way to require everyone to obtain medical coverage, we are finding that there are not enough primary care venues to deliver the care that the new enrollees deserve.

What does affordable mean? How will the costs of coverage be divided between employer, employee, government, and charitable organizations? We all agree that the care ought to be affable—caring, patient-centered with the patient's rights acknowledged.

We seek to provide accurate care that depends upon the best evidence-based medicine, delivered in accordance with recognized guidelines and tailored to the patient's phenotype, as it will emerge from genetic screening and pharmacogenomics.

What our patients really want is advice, good judgment given in a communication style and manner that evinces concern and seeks to advance compliance.

Where will you find yourselves in the decades ahead? You have a right to be proud and excited today. You deserve the best after many years of hard work. You will help to transform the world of health care.

What, by the way, did John decide? He sent me an email message last week:

> "Currently, I'm leaning toward becoming a doctor...
> If I pursue medicine, I want to be creative with my
> practice, to use my influence to influence larger social
> change and close the inequalities that exist with
> accessing health care... This coming year, I want to

pursue things that I think I would be passionate about and take a year off to contemplate the future."[29]

A good decision, it seemed to me.

I would summarize the three components of my title— *Where have all the doctors gone?* They're exercising choice, making commitments to best care, and engaging actively in society's challenges.

Godspeed and again, thanks for inviting me to be with you today.

---

29   In the spring of 2015, I reached John at the University of California, San Francisco, where he was working as a resident in internal medicine. He had graduated from medical school at Baylor College of Medicine and was embarking on a career in global health.

Reflections on Book II

"Too often we underestimate the power of touch, a smile, a kind word, a listening ear, an honest compliment, or the smallest act of caring, all of which have the potential to turn a life around."

– Leo Buscaglia, American author
and professor

My passion for medical education—imparting the best tenets of our practice in our teaching and embracing the art of listening—remains my major focus today. Over the years, I have not seen any deflection among our medical students from the noble goals and career dedication that have always typified the medical profession. Indeed, with the invigorating power of social media and new techniques in learning, I believe the students of today are armed with a capacity to do good beyond what we dreamed of.

I know of no other field of endeavor that offers so much variety and so many exceptional opportunities. Although a bit nostalgic perhaps, I remain convinced that medicine, serving those we are privileged to care for and care about, is the noblest of professions. For those of us who have had the privilege of teaching aspiring young doctors, no one has said it better than the great Canadian physician Sir William Osler in his volume *Aequanimitas*:

"I desire no other epitaph ... than the statement that I taught medical students in the wards, as I regard this

*by far the most useful and important work I have been
called upon to do."*

I still feel a strong commitment to the social needs of our
populations, to the underprivileged, and to the poor among
us, recognizing our responsibility and need to deliver to
everyone the best care possible. For me, a single-payer system
like the one in Canada comes closest to achieving this goal.

How can we improve our healthcare systems? There are no
simple directions or answers to the complexities of healthcare
delivery, insurance systems, and the escalating cost of high-
technology medicine.

This is not the place for a detailed review of factors that
impact health care. I have previously discussed many of them
in *Alfalfa to Ivy*. Here, I will briefly address some of the challenges to and opportunities for the delivery of the best possible medical care over the next decade and beyond.

Communication and Teamwork

Every physician knows the importance of listening, of hearing
what the patient is saying about a health issue. Because it
impacts this communication, I have grown increasingly dis-
mayed by electronic medical recordkeeping. It is the right
thing to do; every patient deserves to have an accurate record
of treatment and a list of medications that are available to
each care provider. These requirements should not abbrevi-
ate or inhibit the direct face-to-face encounter so important
to good patient-physician interactions. A face buried in the

computer with a back to the patient fails in the most elementary of human communication. Documentation requirements associated with the frustrating layers of complexity inherent in medical insurance practices only compound the problem, ultimately stealing time from patients.

Essential to good care will be new requirements for teamwork. Partnership with other members of the healthcare and paramedicine workforce—nurses, social workers, pharmacists, optometrists—will be critical in the future.

I am reminded of what the great humanitarian and physician Francis Weld Peabody said about the physician-patient relationship:

> "The treatment of a disease must be completely impersonal; the treatment of a patient must be completely personal."

But he added:

> "The secret of the care of the patient is in caring for the patient."

Fragmentation

Healthcare delivery is often a fractured experience excessively dependent on specialists rather than primary care doctors. The term *medical home* has emerged as a descriptor to address this concern. A more recent development that contributes to fragmented care is the genesis of healthcare delivery in the commercial sector. In the U.S., we've seen the rapid growth of acute-care walk-in facilities in Walmart and major pharmacy chains.

Cost

Technological advances and newly developed biologic drugs for cancer and rare orphan diseases show enormous promise for effective treatment. Yet today, cancer drugs now receive approval when only shown to extend life for a few months. The value of life versus the cost of drugs will become an increasingly critical bioethical issue. The escalating costs and reimbursement arrangements will require consideration when current practices become financially unsustainable.

We also see continued opposition from our medical-industrial complex to the social value inherent in a sensible plan of health care for all. The escalation in healthcare costs as an increased percentage of the country's gross domestic product would not be negative if the result were job generation and better health for our citizens. But when money and resources disappear into the health care *system*—inefficient, fragmented, specialty driven, corporate-profit motivated—we should worry.

Personalized and Novel Therapies

Reflect on the power of genomics and its transformational impact on medical practice. Still largely theoretical, advances in genetics and epigenetics are inevitably going to create a form of personalized medicine and pharmacogenomics that will increasingly invade all aspects of good care. How will the inexpensive availability of a patient's fully characterized genome sequence lead to new therapies, new warnings of side effects of a drug, and new informed consent rules? Genetic testing for identity of defects in the egg or early embryo will become routinely available, allowing for selection of desirable traits and termination of pregnancy for a genetically abnormal fetus.

Recent medical school graduates will see the average lifespan approach eighty-five to ninety years. Accompanying the great successes we have seen in treatment of hypertension, heart disease, arthritis and the replacement of worn out hips, hearts, and kidneys will come new approaches to care using stem cells, neurotechnology to restore brain function and, in my field, drugs to delay the dreadful progression of Alzheimer's disease, amyotrophic lateral sclerosis, and Parkinson's disease.

Globalized Medicine

Great advances are being made in research and treatments in countries off the beaten path only a short time ago—India, China, Poland, Brazil, and South Africa. New approaches to eradication of polio, as was accomplished with smallpox, together with new means to diminish the still alarming prevalence of malaria and tuberculosis, promise outcomes that will reach millions of people worldwide. I have no doubt we will soon witness improved access to treatments available for HIV/AIDS and for hepatitis C in every country.

Patients may well have traveled recently to other countries, as tourists or, indeed, as patients. We no longer hold the only source of sophisticated care. Travel has brought us closer together, amplifying the possibility of widespread contagious events like SARS, MERS, Ebola, avian flu and now Zika—a virus unique in human history for being transmissible by both mosquitoes and by sexual contact. Physicians will need to be more aware of all of this.

What an exciting century lies ahead. My only regret looking back on the remarkable progress of the past half century is the recognition that the best is yet to come, and I will miss some of it.

Book III. Varia

Introduction to Book III

Among the important considerations for a career in academic medicine is the impact of the growing diversity of our students and the requirement for education of a workforce sensitive to cultural diversity and capable of managing health care inequalities. My experience at UCSF when the regents of the university successfully removed affirmative action as a tool in admission of students and recruitment of staff and faculty had a profound effect on my awareness of the issues involved and in the search for administrative actions that would repair the damage that ensued.

These deliberations led to our adoption of the mission statement at Harvard Medical School in 2000, which states our purpose:

"To create and nurture a diverse community of the best people committed to alleviating human suffering caused by disease."

Finally, I conclude my thoughts on leadership with twelve commandments that I have found important in my leadership experience and which I hope will prove beneficial to my readers.

CHAPTER 11

Convocation, Bethel College

November 10, 1995

Founded in 1887, Bethel College is a private liberal arts college located in North Newton, Kansas. Students major in the traditional liberal arts as well as the sciences. Affiliated with the Mennonite Church, the college graduates approximately 125 students each year.

When I gave this speech in the fall of 1995, I was in my second year as the chancellor of the University of California, San Francisco (UCSF), having previously served as the dean of the UCSF School of Medicine from 1989 to 1993. During my tenure as chancellor, which ended in 1997, a controversy was swirling around the role of affirmative action in admission to, and faculty hiring at, the University of California system.

"I was aware, of course, that this debate was triggered by a lawsuit from a pre-medical student who was denied admission to medical school at UC San Diego. Together with his parents, he had requested and obtained the admission data for the class members accepted in the fall

of 1995. The graph of the data clearly showed that college grade point averages plotted against MCAT (Medical College Admission Test) scores for white students, when compared to those of underrepresented minority students, yielded two overlapping but statistically significant different clusters. The student applicant compared his grades with the minority students selected for admission when he was rejected and concluded that he had been discriminated against. The facts of the case were presented to the regents shortly thereafter, with the outcome a sophisticated media awareness campaign led by Connerly to challenge current admission procedures and press for the total abolition of consideration of race, per se, as a criterion for admission" (Alfalfa to Ivy, 333).

A Case for Affirmative Action

Let me thank you for this opportunity to visit Bethel College—my first trip to Kansas, I might add. In particular, I want to tell you how nice it is to receive such a laudatory introduction. I only wish my parents were here to take notice. My father would have been pleased, and my mother would have believed it!

In addressing a convocation such as this at a church college and before such a sophisticated audience, it is not inappropriate to try to find a biblical text to assure general validity. I would like to offer for your consideration the story of Moses and the flight of the children of Israel from Egyptian captivity. You will recall that in their flight to the Sinai desert, the children of Israel reached the formidable barrier of the Red Sea. Meanwhile, the Egyptians had changed their minds and had begun to pursue them. Moses said, "Bring me my bridge builders." Two members of the tribe were pushed forward, and he said to them, "Build a bridge to the Sinai." They protested in dismay, "We do not have the proper materials. We do not know the depths of the sea. We have not filed an environmental impact report." The sounds of the approaching Egyptians could be heard. Moses said, "Bring me my boat builders." Two more members of the tribe appeared, and he said, "Build us boats to cross the sea." Once again they demurred, saying, "We have nothing for the keel. There are no marine tars. We have no sails or oars."

Now things were truly getting desperate since the Egyptians were close at hand. Moses said, "Bring me my public relations man." One member of the tribe stepped forward, and Moses said, "Abraham, what shall I do?" Abraham thought for a moment and answered, "You go down to the sea and spread your arms. The waters will part, and we can march safely across to the other side. When the Egyptians try to follow, reverse the gesture and the waters will come back and drown them." Moses said, "Come on, you really think that will work?" Abraham replied, "I don't know, but if it does, I can get you three pages in the Old Testament."

My hunch is that most of us here today want to succeed in life. We want to feel good about ourselves; we want to be noticed. We want our lives to be full, and we want others to like us and to encourage us in our endeavors. We want to be treated fairly.

It is from this general perspective that I wish to draw your attention today to the issues of affirmative action and to the changing times of the mid-1990s.

It is now four decades since this country experienced the fervor of the Civil Rights Movement, which began in the 1950s. I remember that time well. While I was completing medical school and doing my postgraduate training in internal medicine and neurology, a number of my friends and relatives from the Mennonite community were marching toward Selma and contributing to civil disobedience activities in the Deep South. The sixties were a decade of mixed hopes and great sadness, with the assassination of President Kennedy in 1963 and the assassinations of both the Reverend Martin Luther King Jr. and Robert Kennedy in 1968.

Government-mandated affirmative action began soon after John Kennedy's death, when President Lyndon Johnson issued Executive Order #11246 in 1965. These requirements were later strengthened by President Nixon.

Affirmative action was in the 1980s, and still is today, a requirement under many federal laws and regulations. It does not require, and in its best application does not mean, quotas or the preference of unqualified over qualified. To quote a recent statement by President Gerhard Casper of Stanford University, our sister university to the south, "Affirmative action is based on the judgment that true equal opportunity means the creation of opportunities for members of historically underrepresented groups to be drawn into various walks of life from which they might otherwise be shut out."

The question that I want to ask this morning is whether we have advanced in our relationships with the minority races of this country, and whether our understanding of racial differences and our tendencies toward racial prejudice have improved or been set back. How do we face the current times in our complex American society, and how do these changing times impact the views and concerns that we as justice- and peace-seeking people should espouse? Have we given women the same opportunities to be leaders in business, in medicine, and in our universities?

Spotlight on California

The events that I have experienced personally over the past few months as a chancellor of one of the University of California campuses, with the rejection by our regents of affirmative action for admission of students, and for hiring of staff and faculty, have put before the leadership of our California educational institutions sobering questions about what we can and cannot do. These events call into question what direction our society is currently taking and what our views ought to be. The issues of race and the questions surrounding justice have been, furthermore, seriously affected by events in our recent history. The Rodney King beating, the trial and acquittal of O.J. Simpson, and the widely broadcast epithets of Detective Mark Fuhrman have all raised the question of whether the races are now more separated than they were in the immediate aftermath of the civil rights legislative actions of the 1960s. Have we suppressed from our consciousness the issue of racism and its impacts on our society?

Let me give you a little history about the issues in California because I think it might provide perspective for our deliberations here today. In the first place, it is important to recognize that California now has more than thirty million people, which exceeds the entire population of Canada. California is really a series of small countries rather than a single state. Depending upon where in the state you seek an opinion, you will find a broad range of liberal versus conservative perspectives. In San Francisco, known probably accurately as the most liberal city in this country, one finds views that are completely different from those of Orange County, near Los Angeles, where the views are strikingly conservative.

The demography of California is also of interest. Of the population of more than thirty million, just over 50% are

Caucasian, and it is estimated that sometime between the years 2000 and 2010,[30] the majority of Californians will be people of color. The state will become a majority of minorities, with the largest percentage—approximately 25%—of the nonwhite population being Latino, Chicano, and Hispanic, and the remainder being African American, Asian, and Native American. I believe, as an educator in a state-supported public university, that we need to address the demography of the state in educating our students to become leaders of the future, assuming that education is a good thing and that leadership will go along with it. Accordingly, I firmly believe that it is important that we educate a proportion of individuals from each of these segments who can take leadership roles in the future.

I believe, as an educator in a state-supported public university, that we need to address the demography of the state in educating our students to become leaders of the future, assuming that education is a good thing and that leadership will go along with it.

Just a few facts about the University of California: We are a nine-campus entity consisting of locations that range from UC Berkeley to UCLA to UC San Francisco. In the case of my own campus, we are unique in having only graduate and professional students. We do not have the privilege

---

30   According to the United States Census Bureau, the population of California was over 37 million and 57.6% Caucasian in the 2010 census.

of involvement in the education of undergraduates. UCSF's schools are limited to the health professions—dentistry, medicine, nursing, and pharmacy. In the graduate arena, we are well known for the excellence of our graduate programs in biochemistry and molecular biology, cell biology, neuroscience, and bioengineering.

Efforts to encourage admission of underrepresented minority students to the university have been ongoing for more than two decades. Success has been mixed. In our medical school, for example, we have succeeded in substantially increasing the number of women who have been admitted to, and graduated with, medical degrees. At present, fully 50% of our medical students are women.

Assessing progress

We have also made progress in the admission of underrepresented minorities, but to accomplish this, we had to take into consideration factors beyond standardized test scores and grade point averages in order to discover and recommend the most highly qualified students from each of the racial groups. Because we have had a mechanism to carefully scrutinize students from underrepresented minorities, we now find ourselves criticized by our board of regents for giving preference to such students. That is, we are accused of admitting students whose test scores are lower than the white applicants who are in the same pool. We have argued that test scores are only one set of criteria to determine the characteristics and qualifications of a future medical doctor and have therefore expanded the criteria for admission to increase the diversity of our students and to make efforts to graduate doctors who will serve the demographic makeup of our society. Our data show that a black physician is more likely to serve inner-city black areas than a white doctor.

> We have argued that test scores are only one set of criteria to determine the characteristics and qualifications of a future medical doctor and have therefore expanded the criteria for admission to increase the diversity of our students and to make efforts to graduate doctors who will serve the demographic makeup of our society.

Our admission procedure is complex. We have six thousand applicants for 141 positions in our medical school. Along with Harvard, we essentially are able to select our class from the positive responses to the admissions process. In other words, we offer only about 225 to 250 students entry to fill the 141 positions. We interview about 600, or one in ten applicants, and if an interview is granted, there is a one in four chance of being admitted. We have a special panel that reviews applicants from underrepresented minorities. Interviews are conducted blind [without knowledge of test scores and grade point average] with two individuals—faculty or students. The subcommittee takes its recommendations to the full committee for review and the final slate of acceptances is drawn from the whole admission committee's recommendations.

The opposition to affirmative action arises from several sources, but fundamentally speaks to the likelihood that we have reached a point in our society where the principles of affirmative action are no longer required to guarantee success on the part of an underrepresented minority individual. The United States gives constitutional guarantees to individuals, not groups. We all agree without, I think, any difference of opinion that the United States ought to be a society of colorblind equal opportunity. In such an ideal society, criteria of success would be determined entirely upon accomplishments and upon those less tangible factors that can be assessed in a personal interview. In such a non-race-conscious society, it would be less important to consider race, ethnicity, gender, and other factors.

The University of California regents have further pointed out that our success in diversification of our campuses has not always led to any real evidence of integration of our students. There have been increased tendencies toward dormitories that specifically designate students of color to live in close contact with other students of the same color. This form of tribalism has been supported by administrators because it may assist students in adjusting to the pressures of university life and in seeking and finding solace and strength from like-minded individuals. But if one is seeking true diversity in a community of students, it is argued, this form of segregation will only emphasize differences and not secure understanding and tolerance.

Politics in the mix

There has emerged a group of effective African American spokespersons who believe firmly that affirmative action must go. It was this scenario that led to the decision of our regents on July 20 of last year to abolish affirmative action. The setting was grossly political. Governor Pete Wilson, a Republican who is a member of the board of regents by virtue of his position, attended this, his first regents' meeting in three years, after having publicly announced some two months earlier his intention to run for president. In June, a month before the regents' meeting, he issued an executive order to remove affirmative action from contracting and hiring practices of state entities. As he built the platform for a presidential bid, he made the abolition of affirmative action one of its major planks. The regents' meeting became a test of his own political force to bring about the first major reversal of the principles that he himself had espoused as mayor of San Diego and as a senator from California.

The meeting was extremely tense. Hundreds of students were there to speak and to protest the abolition. Governor Wilson strong-armed several of the regents, who had been appointed either by him or by George Deukmejian, the previous governor—both Republicans. The issue was settled on a close vote of 15:10 in favor of abolishing affirmative action in three areas: admissions, recruitment of faculty and staff, and in contracting. The fight was led during the six months prior to the vote by an African American regent named Ward Connerly, who had explicitly indicated that he would take on the issue and would fight it to a vote. Let me emphasize that the action was taken against the well-known wishes of the president and chancellors, the faculty senate of the university, and the students. In a recent faculty vote, Berkeley voted 124 to 2 to urge the regents to rescind their actions. That sort of unanimity from Berkeley? Never heard of before.

Regent Connerly declared that as U.S. citizens, we have a warranty that applies to us as individuals, not as members of a group. This warranty includes the presumption that under the government, the standards to which we are held will not be higher for us than for anyone else. Affirmative action was necessary in the 1960s to shock society, but this practice does not make sense for the 1990s. Connerly attested that we cannot give benefits to those who have not suffered at the expense of those who have not caused them harm. He stated that he regrets to say that the University of California is practicing discrimination and that the affirmative action admissions programs at UC are bad public policy. Affirmative action has become a self-fulfilling prophecy by instituting the message that certain people cannot make it. According to Connerly, the University of California had a 3.9% black enrollment statistic in 1980. The current black enrollment at UC is 4.1%, hardly an improvement, in spite of all the affirmative action. Connerly explained that this is so because it's been pounded into people's heads that they cannot achieve. He went on to say that under affirmative action, eligibility and competitive admissibility are different and that the message is for minorities to achieve the bare minimum because there exists an entitlement. He emphasized that affirmative action doesn't encourage hard work; it doesn't encourage people to do their best.

Connerly and certain other black leaders around the country have taken on the issue of affirmative action as an evil that needs expunging from our society. They believe that it has demeaned and lessened the self-image of young black students who either take advantage of it to achieve less than their full capacity—feeling a certain right to admission to our educational institution—or, even when they try with the greatest of energy to succeed, always feel lessened by the pall of affirmative action, which holds that, except for it, they would not occupy the places that they do. One such strong advocate is Shelby Steele, who has written a best seller called *The Content of Our Character: A New Vision of Race in America.*

This year, Californians will vote on a civil rights initiative about whether preferential treatment by sex, race, or ethnic origin should be permitted in any of the activities over which the state of California has any control.

This raises the important issue of whether the centuries of inequality that the black race has suffered can be, or has been, repaired by a mere thirty years of affirmative action—redressing the wrongs of the past by preferences and special opportunity. I shall let you decide on that one. Isn't it curious how we define race? A child of black and white parents is considered black, not white, but a child who has just one American Indian grandparent is considered Indian.

> Isn't it curious how we define race? A child of black and white parents is considered black, not white, but a child who has just one American Indian grandparent is considered Indian.

Calculating the future

What may be the impact of abolishing affirmative action? The sober realization for the highest profile undergraduate campuses of the University of California, namely Berkeley and UCLA, are summed up by the following statistics. The campus at Berkeley, for example, currently admits about 3,600 freshmen each year. If affirmative action is abolished as a methodology to allow students who have demonstrated less achievement in high school as determined by grades and college admission tests, over time, it will become approximately 55% Asian, although Asians account for just 10% to 15% of the state's population. The remainder of students will be white, with few Latinos or African Americans admitted. Do you know that there are more than 4,000 Asian students who apply to Berkeley each year with perfect 4.0 grade point averages? The entire freshman class could legitimately be made up of a single racial group. The numbers for UCLA are slightly less skewed, but are similar. The small number of minorities who would be permitted to enter the University of California would go to some of the lesser known campuses of the system.

It can be argued correctly that the problem lies in the educational system before college, kindergarten through grade 12. The inner-city schools from which the majority of underrepresented minority students graduate are in poverty belts; the facilities are rundown, and the motivation to learn is frequently missing. It is a common experience for bright, motivated black students to be mocked and tormented by their peers for being like "whitey." Many of these inner-school environments are anathema to progress toward graduation. Nevertheless, there are exciting programs of outreach to support these communities sponsored by the university, which we hope will bring increasing numbers to admission eligibility.

The so-called master plan for the University of California, adopted in the 1960s, states that the top 12.5% of students will be eligible for admission, and that enough available positions will be found in the university for any qualified student. At the present time, about 7% to 8% of eligible students enter the University of California. The remainder who attend higher education do so either at community two-year colleges, in the other public colleges that are part of the California state system, or they will go out of state.

Does such an arrangement speak to the public good? Shelby Steele, whom I mentioned a few minutes earlier, has an identical twin. His brother Claude Steele is a professor at Stanford University, and is as left-minded in his views about these matters as Shelby is right wing. Claude Steele, who believes that African Americans suffer from a condition that he has called "stereotype vulnerability," has shown from psychological studies performed on young black students that if they are presented with a test situation in which it is indicated to them that they will be compared to whites, the stereotypic response of heightened vulnerability creates a state of anxiety, in which accomplishment is diminished. When this factor is removed, students will frequently score as well as their Caucasian counterparts.

How should we view this issue? My own belief is that we are not a color-blind society. Racism runs rampant in our thoughts. It is difficult for even the most considerate, informed, and wise individual to avoid stereotypes that have been created by a lifetime of observing the differences in others. In its most negative form, it is racism. In its blander form, it is a presumption of capacity, of talent, or of worth carried on subconsciously and with little voluntary control over it. It is hard to think of a championship basketball team without imagining that the majority of the members will be black. It is difficult to think of a Nobel Prize winner in physics, math, chemistry, or biology without thinking about a white male. The emergence of Colin Powell as a presidential candidate was a fascinating phenomenon. Was it because he is a general? How Americans do love war! Race did not seem to be a factor.

Racism runs rampant in our thoughts. It is difficult for even the most considerate, informed, and wise individual to avoid stereotypes that have been created by a lifetime of observing the differences in others.

My colleague Chuck Young, for twenty-seven years chancellor of UCLA, said recently, "I've never felt that affirmative action is about 'giving an education to poor minority kids.' I've always felt that the benefit of affirmative action is the education it imparts onto the institution." I agree with that.

My colleague Chuck Young, for twenty-seven years chancellor of UCLA, said recently, "I've never felt that affirmative action is about 'giving an education to poor minority kids.' I've always felt that the benefit of affirmative action is the education it imparts onto the institution." I agree with that.

A few weeks ago, our students at UCSF took an hour or two from classes to protest the regents' action. It was a spectacular event. Students from all backgrounds—Vietnamese, Cambodian, Filipino, Chinese, Japanese, Native American, Chicano, Latino, Central American, black, white, Christian, Buddhist, Hindu, Shinto—celebrated UCSF's diversity. I felt very proud to be there.

In my position as chancellor, I have affirmed, in as visible a way as possible with my faculty and students, my firm commitment to the belief that all men and women are created equal, and that we ought to increase our efforts to educate all eligible students and to grant each individual the maximum opportunity. We as white people need to be aware of the irrational feelings and thoughts that emerge from our past experience of stereotypes of races. It is also dangerous to presume that we have made any great degree of progress in our feelings about race over these past thirty years. It seems that different approaches will be required in the future from those taken in the past. I encourage you to think deeply about these issues and to work in whatever way you can to address these needs and to attempt to improve the world in which we live.

Postscript

In November 1996, Californians voted on a public referendum known as Proposition 209. With 54% voting in favor, it passed, amending the state constitution to "prohibit state governmental institutions from considering race, sex, or ethnicity, specifically in the areas of public employment, public contracting, and public education." Now, two decades later, having withstood legal challenges, it remains in effect.

This is an issue that extends well beyond California, and the impact is complex. Enrollment of some minorities in elite public universities has declined, while enrollment of other minorities has increased. For example, today there are fewer black male students enrolled in medical schools in the U.S. than there were in 1978.

I have now had the opportunity to listen to many voices on the issue of affirmative action and to reflect on what has changed in the twenty years since I gave this speech in Kansas. I have been, of course, most aware of and sensitive to the admissions process for Harvard Medical School and earlier for UCSF. I find the "politically correct" atmosphere in these environments to stifle discussion of some of the genuine issues. Some have concluded that the real minority candidates in our admission process are white males.

When we accept at least 20% Asian students—often grouped by this designation despite their ethnic origins in China, Vietnam, Thailand, India, Pakistan, and elsewhere—and make serious efforts to admit 10% to 15% of students of African-American and Latino backgrounds using "acceptable" affirmative action devices, it is quite apparent that the white male is by definition a *minority*. In the admissions process, we have rightly advanced gender balance to equalize men and women. Currently, in many medical schools, women outnumber men by significant numbers.

The argument made most cogently remains that a diverse class creates a desirable, "modern" necessity for the full education of our students. Yet, I would submit that this argument has had little actual impact. Groups of self-identified students too often align that way, particularly in the undergraduate years, at least partly because of encouragement for students to gain strength from each other in a setting familiar with respect to racial background and attitudes.

The greatest challenges lie ahead. Two decades after the events at the University of California and the actions of its regents, the issue remains front and center. A prominent case under review this year (2016) concerns Abigail Fisher versus the University of Texas at Austin. Ms. Fisher's denial of admission to law school precipitated the lawsuit. In 2013, the Supreme Court vacated the decision of the Fifth Circuit Court of Appeals, which had ruled in favor of the university. The high court sent the case back to the lower court for further consideration. In June 2016, the Supreme Court sent notice to the University that it could continue using affirmative action principles in admission policy. The case was decided 5-3 with Chief Justice Roberts voting with a liberal interpretation, carrying the outcome after the death of Justice Antonin Scalia. It appears to me that the logical argument to defend affirmative action in our colleges and universities, driven in part by a moral argument that seeks to assuage the guilt of past discrimination, will in the future carry less weight.[31]

The salient issue that needs to be addressed is the perpetuity of class inferiority brought about by poverty, breakdown of family structure, and poor educational experiences. We ought to improve the methods of education in our elementary and high schools, thereby making candidates better prepared to compete in the academic settings where many students are now inappropriately admitted into studies they have difficulty mastering. We must take a more compassionate and broader view of the effects of socioeconomic disadvantage. At the undergraduate level, as proposed during the California debacle, it may be fairest to admit the top—say 10% to 12%—of the best students from each high school rather than through affirmative action vehicles alone.

---

31   I received a thoughtful summary from my colleague Jules Dienstag, who was Dean for Admissions (1998-2004) and Dean for Medical Education (2005-2014) at HMS. "This issue regarding affirmative action as a public good might benefit from a bit more emphasis. As you mention, the Supreme Court has not been agreeable to the notion that affirmative action was necessary to right previous wrongs (e.g., to exclude past exclusion of underrepresented persons). Instead, the Court has given latitude to educational institutions when the institutions judge that diversity is a worthy educational goal, e.g., to enrich the educational experience of all students, to expose students and faculty to perspectives different from their own, to educate graduates who will be reflective of the society they will join, serve, and lead. The same sentiment has been the persuasive impetus for the military and industry to diversify."

CHAPTER 12

"Diplomacy is the art of getting others to do it your way."
—Holly Smith

Twelve Attributes of Leadership[32]

Looking back on four decades of academic leadership in a number of different roles, I now frequently encounter colleagues who seek advice and counseling on their academic or clinical opportunities. The following is a humble offering of insight into these matters, which I offer with the hope that they will provide useful guidance.

1. If a thing is not worth doing, then it is not worth doing well.

Many of the activities demanded of leaders are a waste of time. Strategic planning, frequently suggested but seldom carried out usefully, is one of them. Often executed at a level of abstraction, it ignores the nuts and bolts of the activity being planned for.

It's not unusual for meetings with faculty to degenerate into gripe sessions, particularly when the topic comes to setting

---

32   Portions of this chapter are revisions from *Alfalfa to Ivy.*

priorities. One can usually decide quickly when an initiative is worth following up on. If not, then kill it before it assumes undeserved importance.

2. Never attribute to malice what can be explained by incompetence.

Among the maxims attributed to Napoleon Bonaparte is, "Never ascribe to malice that which is adequately explained by incompetence." Unknowingly, I have often followed this advice, and incompetence really is so much more common than deliberate malice. I have found it wise to give individuals the benefit of the doubt, often withholding judgment on motivation until more time has passed. I have more often than not found that initial suspicions about malicious behavior are either completely wrong or become more balanced with time.

3. Never tell a lie so you won't need to remember what you said. Be transparent; Openness and honest disclosure spawn integrity.

I have found that a principal reason for the failure of academic leaders is the understandable urge to promise too much when pressed by demands of faculty and department chairs. It is tempting to acquiesce to make the encounter a positive one, but resources are always limited, and choices, sometimes painful, need to be made.

Deferral of response is the key: "I'll see what I can do," and "I'll get back to you on that," or "I need to check with finance and with space commitments we've already made." Here, the

Joseph B Martin

use of the plural is important. As a leader, you have obligations to your team to be certain that they can deliver on what you promise.

Faculty rarely visit except to ask for something. I cannot recall a time when an appointment was made to tell me simply, "You know, you are doing a good job."

4. Rejoice with those who rejoice and weep with those who weep.

At my first meeting with the leaders of Harvard University, hosted by President Neil Rudenstine, who had hired me, one asked rather abruptly about my views on the contentious mergers of the Harvard hospitals and what I intended to do to smooth the waters. I responded instantly with, "I will rejoice with those who rejoice and weep with those who weep," a biblical quotation, which I'd often used but had to later look up because I was uncertain of its origin. (Romans: 2:15.)

Looking back, I'm not certain exactly what I meant, but I expanded on the extraordinary qualities and international reputations of each of the hospitals, adding that I looked forward to learning about what each considered its strengths and weaknesses and where the challenges lay.

As time passed during my tenure as dean at HMS, I came to appreciate a particular trait common to our Harvard setting. Each time a hospital-based faculty member accomplished a great act of public service, clinical advance or research discovery, they were from the Massachusetts General Hospital, from the Children's Hospital, or from the Dana-Farber

Cancer Institute. But when things went amiss and required a public consideration of rebuke or discipline, they were always referred to in the press as "Harvard" faculty. Oh, well!

5. Learn to be generous.

A singularity of leadership success is epitomized in the term "vicarious living." Simply put, it is the joy and satisfaction that accompanies watching the success of others. In an organizational setting, it extends to freely giving credit where credit is due.

There is another aspect to generosity, which is the ability to forgive and forget. Holding a grudge is a powerful disincentive to forward progress. It is impossible to hold a position of leadership without being the recipient of bad news, or news that may reflect unfavorably on your performance or on you as a leader. The source of the derogatory comments may come from important individuals whose roles in subsequent actions are critical. Make an effort to understand the context of the criticism when it is first presented. Harboring negative feelings that surface from the inability to accept the comments for their potential value can lead to persistent counterproductive relationships in future interactions with naysayers.

6. Get out and about.

A key to effective leadership! Leave the office and be visible to the community in ordinary ways. No executive ought to spend more than 60% of his or her time in the office. Experiences there are too confining, and the opportunities to check the pulse and blood pressure of the organization too limited.

Meet with department chairs in their offices, where you have a chance to learn the dynamics of their interactions with staff and observe the inner sanctum of their operations, which can be very revealing. The shift of the power base from your office to theirs can lead to new insights and to greater comfort in relationships. Schedule meetings formally with department faculty in their conference rooms, led by the department chair, where any topic can be placed on the table. These meetings are tremendous opportunities for transparency and a chance to show that no discussion is out of bounds.

Getting away from the office can serve other purposes as well. Attend lectures given by faculty to students to give evidence of the dean's interest in the educational mission. Attend seminars given by faculty, without always needing to sit in the front row. It sends a message of interest in the topic when the dean is seen as a faculty member and not a power figure.

Informal meetings are important. Lunch in the cafeteria with students, faculty, and staff, allows down time spent on ordinary matters, where conversations turn to avocations and outside interests. Regular open-door meetings with small groups of students are useful. For example, after introductions, I would ask them to consider three questions: "Where are you from, and where did you go to college?" "What led to your decision to be a doctor?" And, "What do you expect to be doing in ten years?" The conversation became enlivened with each successive response, and the hour quickly passed with good feelings all around. I sometimes shared my own experience, trying to be honest about the events that had led to my assuming a full-time administrative position.

7. Know when you need to pounce.

This is the ability to sum up a set of circumstances and know when to act, to know when the vectors are aligned to take the next step toward the end game. Judgment requires the ability to know when enough information is in hand to come to the decision. This is the 80/20 rule, to act with 80% of the information in hand without worrying about the other 20%. It includes the ability to define and know when to apply Machiavellian principles to reach a good end for the circumstances.

8. IQ and EQ (emotional quotient): The importance of wisdom.

For me, intelligence encompasses the ability to learn, to remember, to synthesize, to create, to analyze, to differentiate, to classify according to type and condition, to construct new paradigms, to problem solve. IQ implies the ability to innovate, to think outside the box, and to construct new scenarios. But intelligence alone is not sufficient. Individual brilliance may result in earth-shaking concepts, discoveries, and Nobel prizes, but of academic leaders we expect more, or what is commonly called emotional intelligence, or EQ.

Daniel Goleman defines the competencies of EQ as self-awareness, self-management, empathy, and relationship skills. EQ includes sufficient temerity and curiosity to want to understand another's perspective. It includes wanting to learn from another in order to put right one's own views and impressions.

EQ is the ability to understand another's position, to put oneself in the place and context of the whole, to empathize, to understand the impact of group dynamics on the outcome of a situation, to be able to reflect on one's own reactions, to feel and share another's disappointment and pain, to commiserate, and to plan—bearing in mind the effect your action will have on others. Simply put, it is the ability to listen and to recognize what the other person is *really* saying.

9. Don't forget the humor quotient!

IQ and EQ are important, but don't forget HQ, the humor quotient. This is the capacity to see the humor, folly, foible, and absurdity in a situation. It encompasses the ability to use self-deprecation to accomplish an end, to exude a sense of lightness of being, of good cheer and hope. It is the ability to detoxify a situation by humor or self-effacement, to know how to relax the tension with a comment, a story, or a well-told joke. It is the ability to bounce back after an untoward event.

George Valiant, a well-known Harvard professor of psychiatry, has said, "Humor, like hope, permits one to focus upon and to bear what is too terrible to be borne." Quoting another unnamed source, Valiant states, "Humor can be marvelously therapeutic. It can deflate without destroying; it can instruct while it entertains; it saves us from our pretensions;

and it provides an outlet for feeling that expressed another way would be corrosive."

10. Learn to delegate.

It is tempting to try to keep a finger on the pulse of everything that is going on in the orbit of the school, its affiliated hospitals, and the university. Tempting, but impossible! It is impractical to personally try to do everything, so stick to important issues.

Critical to success is a team of trusted, loyal, institutionally minded people. Here are two useful rules. The first is delegate and trust. It is empowering both to the boss and the troops (but inquire from time to time what progress is being made). Second, delegate and encourage. Empower with positive reinforcement for work going well and when it is completed. Give credit when and where credit is due. As Ronald Reagan once supposedly said, "There is no limit to what you can accomplish if you don't care who gets the credit."

11. Be ambitious: Few things happen without ambition.

Successful leadership emerges from ambition to accomplish something. Self-confidence grows from knowing that a course of action is likely to succeed, if implemented openly, promptly, and efficiently.

Anyone aspiring to success enjoys the recognition that comes from wealth, power, prestige, and honor. Fear of failure is a powerful, nearly universal motivating force, and when applied appropriately, it can direct and guide ambition.

Ambition overcomes procrastination, the bane of success, and collects, organizes, and analyzes the available data and takes action, in most cases without remorse. It presses toward doing things well for their own sake, not just for compliments or kudos, but because self-satisfaction in doing the right thing is sufficient.

Second-guessing a decision is a destructive adventure. If a decision seems to have been wrong, it can be adjusted by the next action. Ambition-driven self-confidence ought not to stave off saying sorry when things go wrong, or when a decision is shown to have been wrong. Apologies should be sincere and brief.

12. Know when to quit.

I have always felt that academic (and other) leaders are most effective during the early years of their tenure, and individuals in positions of executive power are most effective in their first decade of work. Hanging on too long becomes a disservice not only to the community but to the incumbent as well. It is no accident that our nation has adopted term limits for governors and presidents. Eight years of service, in two terms of four years each, is a good system. Some might argue that the politics of the president's reelection race every four years limits focus on effective outcomes, but no one argues that a president should serve more than eight years.

Successful academic warriors I have known almost always show great reluctance to step aside from the duties that carried them to the pinnacles of their careers. In every senior academic position that I have held, there have been forerunners who linger on, often with great reluctance to move aside from the limelight. The outcome is often resentment by both parties in the power exchange.

The trappings of an academic leadership position are hard to give up. There is the devoted office staff who shepherd one through the busy academic schedule, arrange talking points for speeches (of which several are often required in a day), make arrangements for meetings with faculty and department chairs and with donors and alumni. Then, overnight, the power of the bully pulpit is transferred to a new incumbent. One's own views and opinions suddenly become less relevant. Free tickets to hockey and baseball games stop, and the contacts and formalities of a daily routine dissipate almost overnight. Most of us choose to be gone, to take a sabbatical to adjust to the dimming of the image, worrying all the while that we will soon be forgotten.

Finally, a word about optimism—and leadership.

I sometimes joke that I suffer from terminal optimism, and it's a fatal case. It is a valuable attribute when you have to fire someone. Leading them from the office, it's important to encourage them that something good is bound to happen to them. As James Reston said, "Stick with the optimists. It's going to be tough enough even if they're right."

Optimism helps when facing trying moments. The worst experiences I've had are visits by a brilliant faculty member, a successful department chair, or an irreplaceable secretary telling me they're leaving. You want to cry but know that would be inappropriate. Quick rebound is critical, allowing for only one or two sleepless nights.

Lead by listening

One of the oldest of the attributes that define leadership that I especially like comes from Plato's *Statesman*, where an analogy is drawn between leaders and weavers. Like a weaver, a good leader brings together people of divergent and sometimes conflicting interests around a common mission, including those who do not necessarily care for each other or for the leader. For the successful leader to weave well in our world, he or she must have the ability and commitment to lead by listening, hence the final aphorism of this guide.

Everyone has a story to tell, and most of us enjoy telling it. Leadership is about listening to the stories others need to tell you; that's the only way to be able to assimilate the conversation and make an informed judgment. Informed judgment requires reflection before any action is taken. A great gift is the ability to listen to all sides in a meeting, synthesize what has been said, and then make a decision that is well-informed and reasonably impartial.

Listen, reflect, and respond.

Further Reading

James, William. *The Varieties of Religious Experience.* BiblioLife, 2009.

Bonhoeffer, Dietrich. *The Cost of Discipleship.* Touchstone, 1995.

Robinson, Bishop John. *Honest to God.* Westminster John Knox Press, 1963.

Lewis, C.S. *Mere Christianity.* Geoffrey Bles, 1952.

Kaufman, Gordon. *In Face of Mystery: A Constructive Theology.* Harvard University Press, 1995.

Sacks, Rabbi Jonathan. *Not in God's Name.* Hodder & Stoughton, 2015.

Carroll, James. *Christ Actually: The Son of God for the Secular Age.* Viking, 2014.

Ludmerer, Kenneth. *Let me Heal.* Oxford University Press, 2014.

Alexander, Michelle. *The New Jim Crow.* The New Press, 2012.

Johnson, Timothy. *Finding God in the Questions*. IVP Books, 2006.

Aslan, Reza. *Zealot: The Life and Times of Jesus of Nazareth*. Random House Trade Paperbacks, 2014.

Carroll, Sean. *The Big Picture: On the Origins of Life, Meaning, and the Universe Itself*. Dutton, 2016.

Acknowledgements

The writing of this set of commentaries and reflections arose from a sorting out of stuff in the office as I anticipated the move to emeritus status approximately two years ago.

I had kept copies of these written documents and, with help, I was able to assemble them into manuscript form.

I am grateful to many people who read and advised on organization and content, challenging me regarding who the book might be written for: Rebecca Pries, Jules Dienstag, Fred Lovejoy, Timothy Johnson, Bob Bryly, Joe Longacre, Garth Bray, Kirk Shisler, Loren Swartzendruber, Ethel Wenger, Michael King, Marian Shantz, and Hamilton Moses. I am especially appreciative of perspectives given on affirmative action by my grandson, Gareth Fowler, who at the time was a senior at Swarthmore College.

It is a special privilege to deliver commencement addresses to students graduating from college, seminary, and medical school, and to join with them in the celebration of their accomplishments. My focus, as in all of my work over the years, is to assist students, staff, and faculty to develop their talents and to advance into positions worthy of their interests and capabilities.

Special thanks to Patricia Cleary for excellent work in editing the manuscript. Thanks also to Beth Beighlie for her creativity in producing the figures for this book. I am grateful to Narjis Corbett for shepherding the process of writing, organizing, and preparing it for publication. The assistance of Sarah Mitchell at FriesenPress was most valuable.

CPSIA information can be obtained
at www.ICGtesting.com
Printed in the USA
LVOW11s1731140717
541097LV00001B/38/P